Over the Water

Essays on Islands

ALSO IN THIS SERIES

At the Pond
In the Garden
In the Kitchen
Dog Hearted
By the River
Freewheeling

Over the Water

Essays on Islands

DAUNT BOOKS

First published in the United Kingdom in 2026 by
Daunt Books
83 Marylebone High Street
London W1U 4QW

1

Copyright © Daunt Books and individual contributors 2026

Island Fragments © Anthony Anaxagorou 2026; Jackfruit and Jaggery © Santanu Bhattacharya 2026; L'isola d'Elba © Octavia Bright 2026; Island Hopping © Nicola Dinan 2026; Take Me Oh Sea! © Ella Frears 2026; As in a Sea Parenthesis © Sinéad Gleeson 2026; You Will Never Touch a Duck-Billed Platypus © Noreen Masud 2026; A Portion of Land and a Cow © Orlaine McDonald 2026; Love Island; Dating Manhattan © Megan Nolan 2026; One Particular Cow © K Patrick 2026; I Picture an Island © Cecile Pin 2026; Isola di San Pietro © Alexandra Pringle; Leaving Avalon © Ralf Webb 2026

The moral right of the contributing authors of this collection to be identified as such is asserted in accordance with the Copyright, Designs and Patents Act of 1988

Every effort has been made to trace copyright holders and to obtain their permission for the use of copyright material. The publisher apologises for any errors or omissions in the above list and would be grateful if notified of any corrections that should be incorporated in future reprints or editions of this book

All rights reserved. No part of this publication may be reproduced, stored in a retrieval system, copied or transmitted, in any form or by any means without the prior written permission from Daunt Books, nor be otherwise circulated in any form of binding or cover other than that in which it is published and without a similar condition being imposed on the subsequent purchaser

A CIP catalogue record for this title
is available from the British Library

ISBN 978-1-917092-68-5

Typeset by Marsha Swan

Printed and bound by TJ Books Limited, Padstow, Cornwall

www.dauntbookspublishing.co.uk

Contents

Island as Escape

Isola di San Pietro | ALEXANDRA PRINGLE | 3

One Particular Cow | K PATRICK | 13

You Will Never Touch a Duck-Billed Platypus | NOREEN MASUD | 23

L'isola d'Elba | OCTAVIA BRIGHT | 35

Island as Metaphor

I Picture an Island | CECILE PIN | 49

Jackfruit and Jaggery | SANTANU BHATTACHARYA | 59

Leaving Avalon | RALF WEBB | 71

Take Me Oh Sea! | ELLA FREARS | 85

CONTENTS

Island as Home

As in a Sea Parenthesis | SINÉAD GLEESON | 99

Island Fragments | ANTHONY ANAXAGOROU | 111

A Portion of Land and a Cow | ORLAINE MCDONALD | 125

Island Hopping | NICOLA DINAN | 137

Love Island; Dating Manhattan | MEGAN NOLAN | 147

About the Contributors | 161

Island as Escape

Isola di San Pietro

ALEXANDRA PRINGLE

I was twenty-two when I slipped my mooring, slid away from my family, from my sense of who I was, who I should become.

In my last months at college, on an orange portable typewriter, I wrote thirty-six letters looking for a job in a language school in Italy. I was offered one in Milan and one in Florence. And so here I was, on a platform in Florence station with a battered canvas suitcase. In the suitcase was a black cotton dress I'd worn to my brother John's wedding, a cream cotton dress, a pair of flared jeans, a cheap sweater and my Biba tee shirts. I had one pair of shoes, green suede cork platforms, and

£200 from my Post Office savings account. It was early September 1975.

The school found a flat near the station for me and their other new teacher. It had central heating, reproduction furniture and what I thought were fake marble floors. I pressed my cheek to the cool floors. They were real. The flat had a balcony looking out onto the Manetti & Roberts factory, the firm that made talcum powder and rosewater in beautiful blue bottles with elaborate labels. When we stepped out the workers waved and shouted, yelling English girls' names. Jane! Emma! Mary! How did they know we were English?

Each day I jumped on the 88 bus from via Corridoni to Piazza del Duomo, ducked into a café for a cappuccino and cake, glanced at that day's lesson before running into the school, housed in an old building on Via dei Servi, the road that winds like a ribbon between the Duomo and Brunelleschi's Piazza della Santissima Annunziata. Outside my classroom window, swallows swooped and dived over the red Florentine rooftops. Inside, I brandished showcards with pictures of men and women, tables and chairs. 'This is a man', '*This* is a woman'.

Summer heat gave way to a sharp chill. Florentines emerged from their homes in their autumn best, grey the colour of that season. Women of all ages conducted their *passeggiate* in grey skirts and sweaters, elegant grey leather boots and handbags. Rain came, day after day, scudding though the piazzas. I went to Upim, a chain of cheap shops, and bought myself a large umbrella in olive green. I soon lost it and bought another covered in tiny flowers. Then came bright, biting cold and fur coats were shaken out of mothballs. I had a brown embroidered Afghan coat from Kensington market that smelt of goat in the rain. Men shouted *gambi di pollo* as I walked by.

Spring arrived and my boyfriend Paco and I drove to the hills to picnic on tomatoes, soft mozzarella and red wine. I raided a fruit tree for its small pink blossoms. Peering down from its branches I saw the farmer looking up at me. He asked if I might consider leaving some for him.

The Italian calendar was a bouquet of saints' days and each long weekend I travelled the spine of Italy in trains. North to Venice and Verona, south to Naples, Amalfi, Positano, Capri. On the train to Naples, I played poker with Sicilian workmen and shared their sandwiches. One delicately traced my ear as I slept stretched out on the seat.

Paco and I went to Elba and it rained and rained. We holed up in a bar filled with soldiers and fruit machines. Shivering, I wore Paco's socks and made a headscarf with his handkerchief for my damp hair. Later we went with my flatmate to another island, Isola di San Pietro in Sardinia, where Paco's family came from, and I fell in love with its rocky wilderness and translucent sea.

We went out on a boat with a friend of Paco's, a coral fisherman. While he harvested at the bottom of the sea, we lay stretched on deck, moving only to jump into the sea to swim. Peering into the water, we saw him lying on his back decompressing before rising to the surface, his net filled with pink coral, the odd fish and a *cazzo di mare* – penis of the sea – that oozed cream liquid onto the deck. The fisherman gave me a little branch of coral from his net; it has been with me these past fifty years. After the day in the sun, in a sensual stupor the four of us lay like sardines on a bed. This was the island where I was to spend that summer.

It was July and my teaching year over. I stood in front of my final class and talked of how lovely they'd been to teach. I saw my students crying at the back. Row by row, more began to cry. How touching this was. Then my eyes began to hurt and tears streamed down my face. There was a riot outside and tear gas seeped through the open windows. As my students left, a communist doctor asked if the song-and-dance routine of my teaching was part of the proscribed course or my own invention.

I was offered a job in Rome and my heart said go. But I knew if I didn't return to London I would stay adrift and I didn't want to be an English expat in Italy. Under Italian law, on leaving my job I was given an extra month's salary. I had cash to spend and six weeks of freedom before me.

Paco gave up his job as a graphic designer to spend the summer with me before he returned to his family in Verona, and I to London and the business of being grown-up. While waiting for him to see out his notice I raided Feltrinelli's English bookshop and lay on my bed reading Saul Bellow, Philip Roth and Vladimir Nabokov.

One weekend, we went on a shopping spree. I bought a handbag glossy as a new conker for my mother and a green snakeskin belt for my sister-in-law. I bought a pair of bright red dungarees and a tiny purple bikini for myself. We stopped at a bar for a glass of Asti Spumante, then with new-found Dutch courage I went to a hairdresser and asked him to cut off my long dark hair. Afterwards, head freshly shorn, in white jeans and a leopard-print halter T-shirt, I stood in bright sunshine in the courtyard of the school as Paco took a photograph.

The next day we got on a train to Livorno, then blagged a lift to the port from a middle-aged couple with a plush car that smelt of leather. Jane Birkin and Serge Gainsbourg came on the radio; as they heaved and breathed 'Je t'aime . . . moi non plus' we squirmed with embarrassment and laughter in the back.

From Livorno we took a ferry to Sardinia. Then a train along the coastline, pink sentinel flamingos etched against the sky. Then another ferry to Isola di San Pietro.

Paco rented a small car and fixed it so the mileage wasn't recorded. We called on his aunt and uncle, ate dry sponge cake and drank a glass of sweet Madeira. His aunt told us her husband's mind was confused, and as he wandered the town he ordered dozens of fresh fish at the market, thinking he was still feeding a growing family. The stallholder handed over piles of glittering fish, knowing that Paco's aunt would bring them back later that day. We picked up the keys to our island home for the summer ahead.

We drove inland to a small, whitewashed stone house, almost a hut, perched on a rocky hillside garlanded with prickly pear and spiky aloes. In the living room, a ladder was propped against a wooden platform on which lay a straw-stuffed mattress, our bed. There was no bathroom, just a plastic basin to wash in and a plastic bidet on metal legs. There was a gas ring burner with a gas *bombola*, a lavatory in a shed outside and two wells. One bucket brought drinking water, the other water to wash with.

In Florence I had received rare expensive phone calls, and letters through the post. But here no-one knew where I was, or how to reach me. I had disappeared.

Each morning we opened our eyes and said, 'Do we want rocks today, or sand? Do we want wind or shelter?' If there was wind to make waves we drove to the sandy bay and jumped up and down in the water, shrieking as the sea ripped off our swimming costumes. If it were calm, we lay spreadeagled on the sand, scarcely moving. But it was the rocks that yielded the greatest pleasures, the grey volcanic rocks that harboured pools of clear water, giving shelter to sea urchins, starfish, mussels and other mysterious sea creatures.

We blew through our snorkels and turned ourselves upside down, diving through the glassy sea to the rocks below. Gently we detached the spiky sea urchins, bringing them up to the surface gasping, holding them cradled in our hands. Back on land, we cut them open with a penknife to reveal the reward – glistening, coral, intimate parts of a sea creature in a dark nest. With our fingers and tongues, we extracted their softness, sliding them into our mouths to catch the burst of flavour.

We picked purple mussels from the rocks, splitting them open, squeezing sharp lemon juice into their orange insides before sucking their fruit into our mouths. Salt was everywhere, on our skin, in our mouths, on our eyelashes, in the whorls of our ears.

In just my bikini bottoms, I lay on the rocks like a loaf in an oven, turning first caramel, then a glowing brown. I wore almost nothing that summer – a thin, saffron, cotton caftan and another in purple silk, a pair of espadrilles, a cotton skirt printed with pineapples and mangoes. I washed our clothes in a plastic basin with cold water and gritty soap powder and pegged them out on a line by the aloes and prickly pear.

Once at a wide sandy beach we watched a German family arrive in an inflatable boat. These were the only non-island people we saw in those weeks.

We were like puppies, so lazy I spoke only English to Paco and he Italian to me, not bothering to trouble with each other's languages. We knew that at summer's end he would return to Verona and me to London, our ways parting, paths diverging. But for now, just for now, we had these slow, sun-filled days.

Neither of us could cook but Paco's aunt taught me to make batter with eggs and flour and water, showing me how to chop *zucchini,* how to dip the pieces in the batter and fry them in olive oil. I cooked pasta with tinned tomatoes, *prezzemolo* and garlic, or fried some bacon and cracked eggs on just-cooked spaghetti to make a carbonara. Sometimes we resorted to frozen fish fingers bought in the local town, their breaded oblongs turning limp in the heat before we got them home.

Some evenings we drove into the local town with its church like a scoop of lemon ice-cream, its terracotta stucco houses and ironwork balconies. We walked down the green-shuttered streets lined with palm trees, gazing at shops with striped canvas awnings. We went to bars for a Campari or a local *aperativo*, or a *gelato* – *zuppa inglese* or a creamy *stracciatella*. If we were feeling extravagant we went to a restaurant by the sea to eat *spaghetti alle vongole* or a *scaloppine al limone*. One night I ate lobster for the very first time, pulling on its pink claws, sucking on the joints of its spidery legs.

On clear days you could see Tunisia in the distance, its proximity in the couscous served in local cafes. I didn't know I would one day, decades later, visit that country with my grown-up son Daniel. He and I drove three hours into the desert, ate tagine and drank wine. A huge moon began to rise. As it climbed into the dark sky, Daniel handed me a small package. I opened it. In the palm of my hand lay a string of glass beads, blue as the Mediterranean, made by Romans 2,000 years ago, its ancient beauty shining in the moonlight. It was midnight, I was seventy years old – and embarking on the business of old age.

At summer's end, Paco and I travelled back to Florence. He took me to the station with the old suitcase I'd arrived with a year before. We embraced again and again, and then one last time, and I climbed onto the train. As it ran through the evening and night, as the cypresses of Italy gave way to the plane trees of France, I cried inconsolably. I shared a couchette with an English woman and her daughter who was half English and half Italian. 'It's best not to marry an Italian,' the woman advised, thinking I was crying for the boyfriend I'd left behind. How could I tell her I was crying for myself, for the me I was leaving behind, the girl with no responsibilities no thought but for the sea and the sun and the rocks and the moment?

I climbed off the train at Victoria station and spotted my brothers and sister-in-law at the end of the platform. I waved wildly but they stared past me. When I stood in front of them their expressions changed.

'Penny wondered if it was you,' Mark said, 'but John and I said it was a Pakistani girl.'

Alexandra Pringle's memoir, *Caravan*, will be published by Canongate in the spring of 2027.

One Particular Cow

K PATRICK

An island will humiliate you. In increments, on the Isle of Lewis, I was humiliated. In retrospect I realise it was, indirectly at least, why I had gone there.

The Isle of Lewis, is a two-hour ferry from the tip of the Isle of Skye, a more famous place. More often I travel from Ullapool where the boat is less likely to be booked out by tourists. In summer, on a good day, you can sit out on the large top deck and watch the land masses retreat and gain at surprising speed; feel glamorous, like you have someone to wave goodbye to, as the engines vibrate. Mountains low-slung, the port small, just a mouthful of green water. If it's still a good day, The Minch won't be too bad. In winter, seasick experts take a corner seat, inside, closest to the largest window. Know how to throw up silently into a bag. I do not understand seasickness. It's hard not to watch. Their eyelids flutter. So much time left on the clock; the crossing has barely begun. They attempt to enter some interior dry land, lips pursed, a patience I can only admire.

I had spent the last few years admitting things to myself. Testosterone yes, but low dose. Top surgery yes, and no drains. Buzz yes, one on top fading to a half. It was hard to understand the exact ratio of decision to action. I'd been in the darkness with it. Right beneath that ledge of being.

The buzzcut was the only admission I could fully commit to. My ex-wife did it. There was plenty she couldn't understand but, during the frightening boredom of lockdown, it was a legible task, an outcome easily understood. Afterwards I thought my head looked good. Was proud, especially, of the widow's peak that had been revealed. The first barber I went to complimented me on it. As I'd waited my turn, I watched to see if the other men closed their eyes as the clippers travelled their skulls. Some did, some didn't. I thought masculinity would be more obvious. All this time.

Traditionally I'm not a crier.

Humiliation can be the willing, living, destruction of self. Sexually, of course this is well-known. The need not only to be reduced to nothingness, but to intensely experience that reduction. I wanted to witness my own annihilation.

I believe I love the Isle of Lewis because I no longer want to be perceived. This desire, to not be perceived, is popular enough to be constantly circulating in different iterations of the same meme. Such desire — to exist without existing — is humiliation's starter pack. On Lewis I'm subsumed by nature, all my looking turns outward, I crawl out from under the ledge. There had to be nothing trans left to think about.

The island offers the gorgeous excuse of metaphor, to keep on detailing the view from my window, the car, the walk. Relentless description. In the beginning it is the pointlessness I'm after. I write the first draft of my second novel. It is gloriously, insanely boring. Mountain after mountain, bird after bird. My body everywhere and nowhere to be found.

The humiliations start small. I had never been bent over by the wind before. For three weeks that first February it did not drop below sixty-five miles an hour, discovering itself, some days, at eighty-seven. And it is like that, wind, always discovering itself. Admiring its own execution, that humming pause before it hurls again, larger and louder. Wind will make a fool of you, worse than any other weather. I did not properly tie down the bins. All three skittered pathetic down the road, the slopes, pinned to the fence of the neighbour's sturdy chicken coop. Each piece of rubbish left on stark display. Not parking against the wind meant the driver's door was wrenched off, left hanging by a hinge. The mechanic, knowing. A malaise, tension in everyday living, the stakes high, tiles peeled from the roof. Every sound a stranger. I sat afraid in my bed; nothing to be done. It dragged wood and stone out from the garden, permitted, inexplicably, a fledging pine to survive.

A walk done every day that involves crossing a plank of wood over a burn, a makeshift bridge. I'm blown off, uncleanly, landing sideways in freezing water. Hip dull against rock. At the sound, two herons, a swan, whatever other water birds, turn to look. There is my body, debased. I feel better.

In Glasgow people say they love my hair, or lack thereof. I can't stand it. But if there's a better affirmation I don't know what it is. Yes, I can't stand the compliments and yet I want them to say it, to notice me constantly. The idea is to be manly, isn't it? To have changed enough that people would understand me anew. In the end it is this that is too much. There is no right thing to say.

ONE PARTICULAR COW

On the island no one mentions it. I am not so much looked through, as past, as if I am a horizon entire. I buy clippers and perform the buzz myself. Drag a dining chair outside. Yesterday they removed the old power lines, the great trunks wobbled and wrenched out of the earth. Locals will come to collect the wood, carving it up with their chainsaws. Three wrens, it's hard to count, are in and out of the low rock wall. Bedsheets whip on the clothesline. Everything remains a distraction and I run the clippers over the top of my ear. The pain is slow enough to arrive that I carry on. Only inside do I see the blood across my cheek, my skull, my neck spread in tracks, like some road spill.

A good friend is a historian of emotions, so I text her, asking for some history on humiliation. A good essay needs history, I ought to be propped up. She replies rapidly.

> Maybe Aristotle on anger would be interesting to start with
> He ties it into being slighted
> What's the relation between slighting and humiliation?
> And also between humiliation and embarrassment
> I think humiliation is Christian
> Acts of penance
> Sack cloths and all that
> To become humbled
> Humiliation and humbling

I like her stream of thought. I laugh, thinking immediately of the polytunnel, its monkish demands. It sits on a croft that has its own small valley. At the heart of the valley is the shape of what I am convinced is a Viking boat, dragged up for some special burial. From the top of the gravel track, sheep disbanded at my footfall, you can see all the way out to Na h-Eileanan Seunta, where puffins have their burrows in the spring. In the sky above the islands, arranged like beastly teeth, a highway of competing birds, silver eels knocked out of beaks. Looking up at them, the thousands of flying forms, feels like the descent into a bodily disorder, maybe a migraine, eyes scattered between wings and feet.

I am not a natural gardener. It does not bring me peace. I like it as an act of penance, even better that I don't know what it is I need punishing for. I make it up as I go along. Humbled, constantly, by small devastations. One day, for no apparent reason, the deer have trampled the rhubarb. Nothing eaten. Hoof marks in the surrounding soil. Each piece of pinkish-green, tough fruit, stamped free from the plant. A rat appears ingeniously, taking small bites from tomatoes and courgettes. It takes weeks to find his hole. I fill it with stones. He makes another. Rabbits at last breach the fence after a year of trying. They fly through the broad beans, so painstakingly patted into layers of sand and rotted kelp, dragged up from the shoreline below. The rabbits are always there when I arrive. Tearing off as I open the gate. Impossible to follow the geometry of their departure, each drawing a different line to an exit.

But I am less interested in the humbling effects of defeat. It is the humiliation that I want, the decision, in the wake of such defeat, to return and try again. Enter handsomely into the inevitability of your own failure.

I don't know how to describe being trans. Living in a world where you are required to describe it, only to have that description denied, proves over and over again a particular catastrophe of language. This is why I don't like nonfiction especially. With every essay I agree to write, I feel cheap with clarification. I become convinced there is no real, loving way to be read. Writing, though, is an act of humiliation. This is where I can understand language again. Why I remain committed to it.

A few months into living on Lewis and I am chased by a cow. A drive to Northton, on Harris, just before the machair is in bloom. Things have turned greener. It's a long time in the car but worth it for the walk, along the steepening coastline, to Rubh' an Teampaill. Time lodged everywhere in strange layers of evidence. Pieces of pottery with differences of a thousand years. Bones, plenty of bones. Shells, too. Easy to be enamoured with what ends up lasting, that the present day would lose their minds over a midden. I laugh again. The walk, as with most on the island, is across a croft. Cows graze, a few sheep. There is fencing but it is casual. The chapel is a ruin like any other. Tiny windows portion out the sea into venerable sizes. All is calm – ish. Down on the white sand is the huge plastic curve of a fish farm, yanked loose in a recent storm. In the wet sand, once I drop onto the beach, are otter paw prints. Recent. The tide has not had its second to wash them away. I've been well-watched.

Back up the slope a few yellow flowers have poked through. I try but can't remember the name. For a different outlook I turn inland on the way back. The cows, plenty of them, are stationary, heads turn with me, an otherwise disinterested audience. I am their poor performer.

An abandoned Land Rover is filled with sacks of animal feed. The red metal body sunk into the ground or the ground rising to meet it. I don't know.

There is, as I continue my approach, a slight change in one particular cow. She moves full-chested as I pass. A belted Galloway. It takes less than a minute for her to break into a hard gallop, finding me at the bottom of a hill. I run, grateful to see a fence ahead, which I jump. She pauses enormously. Evaluates her full bulk. Tears through metal and wood and wire at heavy speed.

ONE PARTICULAR COW

Here I am dying. The thought then light with consciousness, nothing has happened yet, not technically. I am completely alive. We only face each other. Her breath cuts harder through the air, all that flank. I pant, it's ugly.

Behind me is her calf, I realise, seeing the matching ice cream stripe down its torso. The mother charges, I zig zag, exiting through the hole she has made. It's over that easily. They stand beside each other, tails swinging.

In the afterwards of fear, I shake all the way back to the car. Flooded with that nasty electricity. I think already about explaining, how it will be impossible to translate the terror.

At the gate is the crofter in his truck, window wound down. I want to tell him. Angry with it. Alert him to the facts, the cow's motivated evil, the broken fence. But I stay quiet. There is no good sentence. I cannot admit to being chased by his cow. She pawed the ground, she had horns, both of these details true, the beating heart of the moment, and yet, if I was to tell the story, it would only be funny, as if I had designed the cow myself.

Humiliation brings you earnestly back to the body. I double-check my hands. They're still there and hardly mine.

You Will Never Touch a Duck-Billed Platypus

NOREEN MASUD

With thanks to Hobart River Platypus, and the Australasian Network of Modernist Studies

A duck-billed platypus is as good as everyone says. It's better. Its dense flat wet body looks like a sponge you could scrub with; a Beanie Baby you could flop from hand to hand; a bundle you could keep in your pocket and squeeze in your fist, hard; a thick burger you could flip and sizzle. Its feet are delightful afterthoughts. The tail is a robust handle.

You will never touch a duck-billed platypus. They are rare and dangerous and they are not yours.

In 2024, the Australasian Modernist Studies Network paid for me to go to Australia and talk at their conference about the flat landscapes I like so much. I had never been to Australia. I'd assumed I would never go, because of climate change. But when in this life does one get so lucky, to be flown over to Australia? Also, Australia was flat, I was told: the oldest and flattest continent. I weakened. My first flight in five years. I'd stay on for a bit, after the conference; make it worth my while.

Unfortunately, the conference would be held in Hobart, on the island of Tasmania: one of the only not-flat bits in Australia. But what Tasmania lacked in flatness it made up for in platypus. If you ever saw a platypus – if you were lucky enough to glimpse one of these elusive, quick, endangered creatures – it would probably be in Tasmania.

Hobart is a capital city, full of people and filth, and therefore not a good place to see platypus. A good place to see platypus in Tasmania is the northwest of the island: Deloraine or Latrobe. I looked at Deloraine and Latrobe on the map and rapidly determined that I would not be able to get there. I could not drive. A bus would take almost six hours. This island's size confounded me: a little snip off the edge of Australia was somehow vast, the same size as Ireland. And then there was the whole of the rest of the country, hovering just behind, almost too big to see.

I had learned, in my life, to manage what I hoped for. Because mostly one is disappointed, or one doesn't

YOU WILL NEVER TOUCH A DUCK-BILLED PLATYPUS

get what one wants. So, I would stay in Hobart, and not-seeing-a-platypus would be a kind of game I played with myself. Because there was a chance – a faint one – of seeing a platypus in Hobart. Just like there's a faint chance, always, of seeing a badger or a red squirrel in Scotland. It'd be a once-in-a-lifetime thing, probably, but theoretically the chance is there. So, you go and you have a look, for the fun of it, and the going is the point.

I did my research, for this not-seeing-a-platypus game. A gentle website called Hobart Rivulet Platypus played resignedly along with me. The chances were slim, it emphasised. Really the northwest was a much better shout. But if you are 'stuck on the idea' of trying to see a platypus in Hobart (I was), then, it suggested, you could walk along Hobart Rivulet from the town centre. There were platypus in there – a population of 'low to mid double-digits', they estimated. And if not, there was a big mural of a platypus on Wynyard Street. 'You'll definitely see a HUGE platypus there!'

Squinting in and out with jet lag, shooting up into the top of my head and down into my fingertips twice a minute, I told everyone at the conference about my fool's errand. Mostly people laughed. 'I've lived here all my life,' one person said, 'and I've never seen a platypus. But good luck!' By day two of the conference, I had repeated my plan to walk down Hobart Rivulet and look for platypus widely enough that I really had no choice but to set off, once the day wrapped up, even though it was clouding over and starting to rain. I had nothing warm to wear except the navy-blue extra-large men's hoodie I'd bought to swaddle myself on the flight over. I had heard Australia would be hot, in December – dangerously, frighteningly hot – and within that context I had simply not been able to interpret warnings that Tasmania would be cooler than the mainland. Cloud billowed up over Mount Wellington, piecing together around too-thin cracks of light in the sky.

I started from the centre of town, as instructed, past the shops with their shutters rattling down, being locked by employees wearing coats over their uniforms. How long was the rivulet, I wondered? How long should I allow, walking along, before I felt that I'd given it a really good go and could gracefully give up? Two hours would be plenty. I'd see the rivulet. I'd have a walk. I'd tell my mum about it later.

The city gave out on to something like a park, with water running past, and Tasmanian nativehens with chunky yellow beaks running faster as they spotted me. The platypus information signs began, busily managing my expectations downwards with a kind of scrubbing motion. Things to look out for: ripples in the water (the platypus might just have dived); bubbles (it might be underwater, blowing bubbles); a kind of V-shaped disturbance (it might be swimming along). Look carefully at the tail: people often think they've seen a platypus when really it's a rakali, a water-rat. Long tail, rakali; short round tail, platypus.

On a day like this when you are tired and travelling alone, and colder than you would like to be, you have to hang your good humour up on a high rack to keep yourself upright. It was the case — I allowed the thought in like a puff of exhaled air — that platypus were hard to see at the best of times. And I, in particular, was not good at seeing things. I noticed the things I wasn't meant to notice and could never find what I was looking for. When I move my gaze through a space, it splinters in sixty directions, along the lines and outlines of everything it encounters. So whenever I went birdwatching with my housemate, he would see the chiffchaff or the song thrush instantly, and though he'd arrange me patiently and point as carefully as he could, still I wouldn't be able to see it, even if it was right in front of me, even if it was definitely there. What chance platypus?

It was raining. Not hard but persistently, like a child grizzling, making my hoodie wet and heavy. And for some reason there were crowds and crowds of runners, appearing in twos and threes, so I had to step off the path or else experience the unsettling feeling of being surrounded, engulfed. Their healthy white arms, with their red fatty flush, and the pink Lycra of their vest straps holding it all in place. I like the look of ordinary thick flesh, neither fat nor thin; it seems unsure of how to move, guided by neither gravity nor bone. There's something ethereal about it, like it's been paused. It's buoyant and still on the body but not sure how, and the blood moving through it seems disconnected, non-intrinsic.

The runners kept coming, so jaunty and chatty, two with two huge dogs, and I in my huge aeroplane hoodie, my hands swallowed up and lagged with warm rain, shambling and teetering on my bad knee. All the old feelings: of being always the one forced to move, to compromise, because I was too weak or ugly or brown to be given way to. I felt unhappiness paddling around in my shallows, looking up at me, debating whether to come closer. It was too early in the trip to get unhappy. It could not be allowed. I stepped on to the bank, as runners flooded past, and stared at the water, unseeing.

If nothing else, I thought, I was next to a river in which, somewhere, there was a platypus. Such a thing was almost unimaginable. That I had wound up here, on an island where these creatures swivelled their flat beanbag bodies through the water. The water was solid and the platypus was solid and so was I. I stayed still and imagined my body reaching out across the city, all my skin becoming alive and present in a world where a platypus existed, a finite distance away from where I stood now, occupying the same sphere as me. If I stood in the water we would be in the same water. I was already wet.

I stayed on the bank, trailing along, dipping on and off the concrete path as the runners turned back and returned the way they came (I recognised the two huge dogs). And then, for a while, it was just me. The pockmarked water stilled on the path.

Gradually, I became aware of people behind me, pausing and staring into the rivulet, walking and pausing again. A girl with pink hair, and a couple. The man was very tall, and when he looked into the water it was with the air of a man who loves his girlfriend, who would not be there otherwise. He didn't care about the platypus, I thought. His girlfriend cares about it, and he loves her, and so he is here to look too. And that's sweeter than loving the platypus. I slowed down to listen to their conversation, gazing studiously out at the water, with all my ears open.

'He said it was down here the other day.'

'Rare though. Odd.'

'We'll keep going. We're not at the dam yet.'

I slowed still further, allowed them to pass me, then trailed them at a safe distance.

Near a playground was a big sticker with LARILA'S LEGACY printed on it in pink. Larila was a platypus,

I learned, who had died when twine wrapped round her. Big blue tears streamed out of the sky, over Larila's face. Posters urged me to Seize It, Snip It, Bin It – to cut any loops I found in thrown-away litter, and make sure they were tucked safely into a bin.

Then we were at the dam. I'd worried I wouldn't know the dam when I saw it. But it couldn't have been clearer. Concrete ramps and metal railings all surrounding smooth water, almost untouched, except in one place. When I saw the platypus I waited, and felt nothing, because it had happened so promptly, so improbably. I'd reached the dam and there it was, too easy, like a children's book.

The platypus dived, and little circles of bubbles rose, concentric circles spread out, luxuriously, pensively, right to the edges of the water. It dived, and came up, and dived. I remembered the sign, about how water rats were often mistaken for platypus. And I looked very carefully at the tail, as it dived. And it was round. Still, I waited, for something to happen, inside me or outside.

The trio were pointing and taking photos, and doing their best to keep their shouts of joy muffled. Perhaps it was a platypus. I folded myself right over the metal rail to burn a line of wet into my hoodie, and stared, and stared. Late sun glinted off the fur, so that head and body were two little faded ovals of light in the water, with the suggestion of a third whenever the tail surfaced. It dived.

I slid myself in next to the girl with pink hair. She was a photographer, with a good camera; she showed me her video. I could see that it was a platypus, that the thing that looked like a beak was really a beak. I looked back at the water and started to believe it.

The platypus stayed long after everyone else had left, and night was starting to fall, and my hoodie was soaking wet. It tweaked itself, beak first, through the water; it floated and glinted in the sharp paling light. It parked itself on the concrete ramp, far away, and seemed to scratch itself luxuriously as a bird does, or a cat, humping itself up and scratching again. Then it slid back into the water. Still my mind shimmered, unable to store and trust what I was seeing.

This was perhaps the most amazing, unasked for, unlooked-for thing that had ever happened to me, and it was a little gleaming smudge, a double-smudge, in a city dam, floating on the water like a wisp of twisted cling-film, like God's thumbprint on the hasty world. And I thought about how all the tremendous things might just be smudges at the edges of our mind, tiny, like Icarus's legs disappearing into the water, barely a splash, only two marked ripples spreading out and out.

The platypus was still there when I left. In the end it was I who left it there.

When I headed back, dusk was really falling. Now that I was walking away the confusion was turning into joy. I put on my headphones and turned Sinéad O'Connor up.

I stared at Mount Wellington, at the washed-blue sky with afterthought clouds. And I wondered how I had won this impossible lottery.

And then there was a wallaby hopping across the path.

Things got blurred in my mind at that point. The next thing I remember, I was standing and looking up at the bank by the road and it was covered with miraculous animals, ones I'd only ever seen in books, metres away from me, nibbling and quivering. Grey wallabies, rustling, and other beachball creatures, the colour of red squirrels, what I would later discover were pademelons. I was holding my head very straight and still and trying not to breathe, and a huge grey wallaby was staring straight at me, trying to work out what my game was. The other wallabies and pademelons were eating, because the big guy was on it, they could relax. I moved very slowly forward. Down by a little tree, by the road, a little pademelon froze mid-mouthful and stared at me. A wallaby got down on all fours and crawled, very slowly; it could hop in big motor swings and crawl with the tiny, finicky precision of a spider. It was a machine and a rubber band and an insect.

YOU WILL NEVER TOUCH A DUCK-BILLED PLATYPUS

I was outnumbered and night was falling and they were things I had never seen before and would never see again, all looking at me and eating. No one was there but me, and these creatures, out on their mountainside, in the suburbs of this city, on the edge of this huge tiny island.

When I got back to the hotel I couldn't feel my hands, and it was half past nine at night and everything was shut, and nothing to eat: the kind concierge brought me some chips and mayonnaise and I ate clumsily, drunk with cold.

Later in the trip I would see little penguins hurrying out of the sea, and koalas asleep with their hands curled round branches like a newborn with a finger; kangaroos leaning back on their tails and black swans on glittering sea and a baby echidna padding importantly down a dirt track. But nothing surpassed the platypus day. Because once in a while you put out your hand wryly, half-jokingly, preparing to snatch it back almost at once, and the world quite unexpectedly heaps it until it's overflowing, with more than you could ever have thought or dreamed of. Once is enough, really. Once changes everything. Afterwards, whenever you put your hand out, a little part of you will mean it.

L'isola d'Elba

OCTAVIA BRIGHT

L'isola d'Elba lies off the coast of Tuscany and is shaped like a child's drawing of a fish. You have to squint a bit, but if you do you can see a passable dorsal fin, and the two longer fins of the tail. This was one of many charming things I learned about the island last summer when Vic and I looked it up on Google maps. We were trying to decide where to escape to – me from London, she from Milan. All I knew about Elba was that Napoleon had once been exiled there, so it stood as a steely fortress in my mind, but we were swayed by the caption on our screens: Elba, island with beaches & a tropical vibe. *Let's go somewhere we can have a little flat within walking distance of the sea,* Vic texted, *a small place where we can potter easily but not be distracted by too many things so we can get some good work done too.*

We weren't looking for exile so much as respite, Vic from a recent break-up and the unnerving pitch of the Milanese summer heat, and me from the relentless demands of accountants and lawyers as I tried to sort out my mother's finances. She'd had a stroke the previous year and responsibility for her affairs now fell to me. It was a sharp yank up a rung on the ladder of familial duty, which I'd been marooned on for years (willingly, unwillingly, and everything in between) during my father's decline from Alzheimer's, and which, when he died, I thought I might be able to climb off for a while. But the death of one parent is so often followed by the collapse of the other, and so, in spite of my relative innumeracy, I was now intimate with the state pension system, and my life was a chaos of cardboard folders in muted colours labelled things like IMPORTANT DOCUMENTS and DO NOT LOSE THESE!!! and DEATH AND TAXES. I regularly spent hours (*hours*) on hold to HMRC staring down the barrel of my own mortality to the loop of an insistent, peppy beat. I was behind on everything, including my own work. A few days with Vic in a little flat within walking distance of the sea on an island shaped like a drawing of a fish was exactly the tonic I needed.

L'ISOLA D'ELBA

I land in Milan well after sundown and step out into thick heat, the kind that clots the blood and puts the body on high alert. When I arrive at Vic's, her kitchen smells of roasted vegetables which we eat with stracciatella while sweat runs down our backs. We flee the city at dawn, relieved to know that within a few hours we'll reach the port and fresher air. I glance at Vic in the driver's seat as we pick up speed, her sunny face poised on the brink of a grin. I love how well I know this face, how I can read the minute shifts in expression when different moods move across it. I have known Vic for almost twenty years – nearly half our lives. Looking at her now it's impossible that we're old enough for this to be true. She doesn't look it and I don't feel it, especially not today, riding shotgun on an Italian motorway, gleefully regressing with every mile that separates me from the contents of those folders back home. But it's also a trick of old friendships, this bending of time. Especially if your daily closeness is interrupted by geography. Spending islands of time with Vic plugs me back into a version of myself that I miss when I'm chasing up lawyers or wading through paperwork. It was closer to the surface when we first met at university, the bit of me that's quick to laugh and to forgive, that's optimistic and indulgent, maybe a little deluded, living freely in the present tense without anxiety about what might follow. It's the part that prioritises pleasure and has never bought adult nappies or arranged a funeral. I roll down the window and gusts of air snatch lyrics from the speakers. For the first time in a long time I am nowhere else but here.

Another motorway and a ferry crossing later, we pull up outside our rental on the outskirts of Portoferraio, the island's largest city, which sits at the tip of Elba's dorsal fin. *Ciao ragazze!* calls our landlord for the week as he steps out onto the patio. Eduardo is a tall man with a caramel tan and delight in his eyes at the sight of us. We indulge his gentle flirtation as he shows us around. I don't mind it, but I know that if a man greeted us with 'hey girls' in English I'd bristle. Where 'girl' lands as infantilising and entirely inappropriate for a woman of thirty-seven, *ragazza* has a strength to it in its guttural 'g' and the percussive double 'z'. To my Anglophone ear it sounds confident, rebellious, a little bit wild. It feels like a conduit to the part of myself that time with Vic revives, a spell with the power to summon it. Which, I realise now, is another reason I'm here – there's something I've been needing to ask this *ragazza* part of myself, and I couldn't do it from home.

Some questions are hard to answer from within the experience of your real, actual life. You can't tell the shape of the mainland while you're standing on it – you have to climb a mountain or set sail for an island so you can get a different view. For some time I had been trying and failing to work out whether or not I wanted to have a child. I was exhausted by it, this question I thought I'd answered definitively years ago. Since my early twenties I had known with comforting certainty that parenthood (let alone motherhood, that freighted institution) was not for me. It was liberating to know this, to be free from a desire I understood to be a complicated burden for many of my friends. Vic felt the same, and as people in our lives began to bring babies into the world we enjoyed our new roles as exciting aunties without wavering one bit in our thinking.

But in the last handful of years my certainty had been replaced by a nebulous yet consistent curiosity, and the question expanded in my psyche until it was squatting there, uninvited, taking up an unreasonable amount of space. It's because your father died and your mother is sick, I would tell myself, it's just mortality panic. Or, it's just your hormones, your age, the pressure of social convention. It's just because you're flattered that anyone – no, that John in particular – wants to have a child with you. Or just because you like some of your friends' kids, or because some writers and artists you admire have said it deepened their practice, their commitment to life. It's just because you like emotional intimacy, because you enjoy intense relationships, because you're curious about a different way to experience time. Or maybe it's just because you're compelled by colossal, irreversible change, because you like to take risks, and on some level you just really want to blow up your life. 'Just', what a pernicious little qualifier. Remove it from each of those statements and what you're left with is a list of reasons, simple as that.

My mind was clearly changing and I was struggling to let it. Attached to my old way of thinking I feared letting that younger, more certain version of myself down. What if she was right? What if I had a baby and regretted it? I was afraid of locking my inner *ragazza* into a life she might grow to hate. And I needed her to somehow reassure me.

Each morning on Elba we wake early and do five sun salutations looking out over the balcony and its view of the sea. We sleep on twin single beds in the living room with all the windows open to encourage the breeze. Our daily alarm is birdsong and a beam of sunshine that falls each morning like a spotlight across my breasts, its heat already intense by 6 a.m. The floor tiles are a bright turquoise and so shiny that we dip towards our own reflections as we flow lazily through the asanas, pleased to be honouring our good intentions for once. This is my favourite way to live, early to rise with the hot bright morning, sleeping lightly under only a sheet. I want to hold onto this lightness, my purpose clear, barely any objects weighing me down. We have deadlines to meet so we set ourselves up on the balcony with coffees brewed on the old ceramic stove then poured over ice, and chase the shade as it sweeps slowly across the terrace. In the heat of the afternoon we close our working day with a swim. By the second day we feel like this has always been our life.

Where are you guys at with the baby stuff? Vic asks on day three. We're lying on loungers on a stony beach under striped umbrella shades the same bright blue as the sky, eating coconut husks sold to us by a friendly Sicilian man who combs the shore with a bucket full of ice, palm fronds and tropical fruits. Nowhere new, I tell her, lighting a cigarette. I inhale deeply and let the nicotine buzz silence the voice in my head that scolds me for it, that reminds me of my genetics, my mother's heart disease and its relationship to her lifelong dedication to smoking. Besides, island cigarettes don't count. They happen offstage from reality. I still don't know what I want, but we're not using contraception, I say, so maybe my body will just make the call. Hearing myself I think, how silly, as if your body and mind are separate things. Sounds like you've made your decision to me, says Vic.

On our last night we walk down to our local beach bar for dinner. It's on the wrong side of the island to see the sunset but we don't mind. There's a band set up at the front, a guitarist and a singer. She has thick red curls and a silver starfish necklace that hangs low and rests on the mound of her stomach. She is big, beautiful, with a playful lilt to her voice. A powerful light attached to the back of the guitarist's wheelchair beams a column of white onto the music stand in front of them, and he plucks the strings of his instrument, coaxing sound out with a delicate sort of command. He lets her take the limelight but it's really about them both, the synergy between them. Mosquitoes fuss around them like a cloud of tiny groupies.

The guy who runs this place is small and has one of those straggly ponytails that gathers beneath a bald patch. It hangs, limp, against the faded floral pattern of his shirt. He wants to flirt and show off his English but we reject his offer to translate the menu and it irritates him. Vic's Italian is fluent and mine is fine, enough to get by, but it's more a response to his need for our attention – grasping, desperate, with anger just beneath the surface. We may be the customers but it's clear he expects something from us as well. For the rest of the night he ignores us and knocks pointedly into the bag that hangs off the back of my chair.

Vic lights a cigarette and offers one to me but I don't notice because I can't take my eyes off an older couple sitting a few tables away. They're all sinew and bone, with lupine features. They could almost be siblings if it weren't for the body language that makes it clear they're together. Both wear their hair slicked back into tight little knots, cheekbones high, triangular faces keen with intensity. They smoke ceaselessly apart from when their seafood arrives. In between cracking crab legs, the woman moves her hands to the music like a flamenco dancer, wrists supple, long fingers striking poses to the beat. Her white halter top emphasises the breadth of her shoulders and her muscular frame. It's hard to tell how old they are but they've got to be pushing at least sixty. Whenever they rise from their seats they move through the restaurant as though we are all guests at a party they're throwing, smiling graciously, aware that their presence is an offering to the rest of us. In the woman's hand is a brown and gold fan that she wields like an empress. Finally I notice Vic's outstretched arm, take the cigarette and light it. You know, if you don't have a kid that could be you and John one day, she says, following my gaze. I try to imagine it, the two of us old but fit and strong, with no one to take care of but each other. Freer than I am now. Maybe living on an island, swimming by day and dancing by night.

And then they are gliding towards us, each carrying a tumbler of rich brown liqueur. *Ciao ragazze*, says the woman in a rasping voice as she pulls up a chair. They ask where we're from and we get talking in a genial mix of Italian and English. They're from Corsica, and come to Elba every year for a month in the summer because, says the man, it is paradise, drawing the phrase out so each syllable lands with the same tempo: be-cause-it-is-par-a-dise. The rest of the time they work as croupiers in casinos all over. Have a drink with us, says the woman, go on, and when I decline she flies into a pretend rage, *brutta*, she calls me, again and again. And then it starts to seem less pretend, like she's actually quite angry. I consider trying to explain it's not because I'm square but I have no idea how to say recovering alcoholic in Italian so I let it go, play along, allow her to misread me.

It doesn't take long for their words to start slurring into one another and for their conversation to become repetitive. They want us to go with them to Da Giannino, a nightclub up in the hills that doesn't open until midnight. I look it up on my phone and it's right where the fish's gills would be, if the fish were a fish and not an island. Another night, we tell them, thinking of our deadlines and our sun salutations and our long journey home. When the woman flashes us her thong for the fourth time we start to wish they would leave us alone. Vic catches my eye and grins. Never meet your heroes, she whispers.

Later, once we've settled into our twin beds for the last time, I look over at Vic in the moonlight. Her face is soft, ready for sleep. In it I see a trace of every version of her I have known, and the blueprint for all she's yet to become. I don't know it then but by the end of the month I'll be pregnant. When I come to write this my unborn daughter will be kicking softly against my ribs. This trip to Elba will feel like a pause on the threshold between two different lives, and I'll still feel the pull of the other one. But as long as there's Vic there's a bridge between the two — that's something we'll always be able give one another. For now, we are two islands in a turquoise lagoon. We are an archipelago. Outside, the Tyrrhenian Sea laps gently against old stones and Eduardo's red hibiscus flowers close up for the night. I can hear distant strains of music bounce between the hills and imagine it's coming from Da Giannino's. When sleep finally takes me I dream of our croupier friends dancing under a bright sunrise, their bodies liquid and golden.

Island as Metaphor

I Picture an Island

CECILE PIN

There is the warmth of the water, the fish swarming around me – I feel them brushing against my legs, nipping my ankles. My feet sink into the sand, warm as well, and then there are the waves; they rise high with a thunderous noise, and for a brief moment I fear them engulfing me entirely; but when they fall, it is not with a crash but with grace. When they retreat back in the water, their noise lulls me.

I sense all this – the sea, the fish, the glow of the sun as it sets – but I'm not there. Instead, I'm in a small Parisian office, on a grey and cold day, eyes closed. I am eighteen years old.

I'd been to Phuket with my family the previous summer. We went after I finished my Baccalauréat – France's A-level equivalent – and in the autumn, I was due to move to an island of my own: Great Britain, to study Philosophy at University College London. That summer was a brief, limbo-like respite in an otherwise eventful year, that held in its palm all the possibilities that awaited me in my new life abroad. It was, perhaps paradoxically so, a time in which I felt both a great sense of tranquillity – from having passed my exams – and a dizzying sense of freedom.

On our first day in Phuket, my sisters, my parents and I went to visit Phuket Town. We made our way to the Old Town, seeking the shade wherever we walked, not yet used to the heat. The streets were lined with pastel-coloured, two or three storey houses reminiscent of Portuguese architecture but with traditional Chinese roofs, cladded in curved tiles. We were all struck by their uniqueness, even daring to come out of the cool shadows to admire them more closely. I took many photos, which I still have. Years after that summer, I'd learn that Phuket had been a centre for tin-mining, back in the sixteenth century. Its riches attracted many Portuguese settlers, and to build their houses and establishments they employed Chinese workers. The two styles weaved together: the colourful walls, the curved tiles, carved doors and arched windows – would eventually create the Old Town.

I PICTURE AN ISLAND

Soon after moving to London, I realised that university and building a life abroad weren't quite the fun, plain-sailing experiences I had dreamt of. Which is why I found myself, only a few months later, in my Parisian therapist's office once more. I was back in a cycle of exams, adrift in a place that was neither home nor strange, using a language that was not quite mine. On the wall to my right, I fixed my eyes on the prestigious psychology diplomas, framed and hanging just above the small, mostly empty bookshelf. In the corner of the room, behind Madame Z there was a fern almost as tall as her. She looked at me patiently, her head resting on her fists waiting for me to speak. In between sobs, shifting uncomfortably in my chair, I told her that this move had all been a mistake; that I'd been delusional, thinking I could make it in London on my own.

'Let's do an exercise,' Madame Z said in response. She asked me to close my eyes and to pull my shoulders back, to picture a place where I had felt calm and joyful. She asked me to take a deep breath and to release it slowly through my lips, while picturing this happy place. And so, I pictured myself that summer before I'd moved to London, on the island of Phuket, knee-deep in the sea, a moment that felt both entirely cliché and personal. I remembered my sisters and I taking photos of one another on the beach, the shade of the palm trees and drinking straight from the coconut with a straw. I remembered delicious food, tom yum, seafood as fresh as could be, and the kindness of the locals. And on top of all these things, I remembered vividly the emotions that accompanied them: a sense of joy and of utter relief, at being done with school, at my awkward teenage years coming to a close, mixed with the anticipation for all that was to come.

The only person in my family who had been to Thailand before this trip was my mother. As a teenager, she had spent almost a year there at a refugee camp, having left Vietnam with her family after the end of the war. Like many families of Boat People, only a small fraction made it safely to shore. My mother's parents and younger siblings were lost at sea, in circumstances that remain unknown. This wasn't something that we talked about openly, growing up – and for a long time, this part of my heritage was shrouded in mystery.

Immediately after doing the breathing exercise in Madame Z's office, I felt calmer. 'Every time you start feeling overwhelmed, just take a deep breath, and picture this place,' she told me. I thanked her and left her office, and the next day I took the Eurostar back to London, still unsure about everything that awaited me there.

I PICTURE AN ISLAND

It wasn't until the pandemic, years after university and exams and Madame Z, that I began researching more about my Vietnamese heritage, in a project that would eventually become my first novel. I read page after page of testimonies and newspaper articles on the Vietnamese Boat people; national archives, everything that I could get my hands on in the confines of my own home. And it was during one of those research sprees, in the late hours of the night, that I learnt of Koh Kra, a small island in the Gulf of Thailand. During the late 1970s, a group of fishermen would trap and strand female Vietnamese refugees there, to assault them or worse: victims were in the hundreds. But something else that dazed me, besides the horrors of the island, and the realisation that perhaps my own family had been one of its victims – was learning that the island had now become a popular tourist spot for snorkelling.

Right after reading about Koh Kra, I pictured myself in Phuket once more. On our second day we had gone to a little islet, not too far from shore, for snorkelling. I'd never done it before, and I recalled my excitement at seeing the ocean anew. I remembered the nauseous boat ride there; finding the mask stifling; swallowing heaps of salt water and the eerie quietness as soon as I submerged my head; the little clownfish and parrotfish swimming, the rays of the sun reflecting onto their silver scales. I remembered all of this, sat at my desk in the late hours of the night during lockdown, and then I looked up at my computer screen, detailing the horrors of Koh Kra – and I thought of how easily my snorkelling island could have been the same place.

When I imagined my island in Madame Z's office, I saw it as this: a quiet, secluded place, far from home and my everyday life. It does not escape me how quickly those lavish qualities can become sources of horror. I thought about the Caribbean Islands: their fertile soil, the climate, the strategic location that attracted many European settlers there in the seventeenth century. Almost all of the islands were colonised, and more than three million Africans were brought to them and enslaved. Like Phuket (which was never colonised), the Caribbean's biggest industry is now tourism.

I thought about how islands, microcosms of their times, evolve. I thought about how they hold within themselves both the makings of hell and paradise, and how to make sense of those contradictions. When I learnt about Koh Kra, I could no longer picture Phuket as my happy place. I felt the warmth of the water, and I wondered if it had been that same body of water that had swallowed my ancestors. I felt my feet sink down in the hot sand, and I imagined how hard it must have been to escape in it. I saw the waves rising, and I imagined them engulfing a small, tattered boat, carrying mothers and infants.

In 2017, Fyre Festival – a luxurious music festival, set to take place on the Bahamian Island of Great Exuma – went viral. Ads for Fyre showed a dozen supermodels arriving by private plane, running along the shore and swimming in the sea, playing with the local pigs. When the five-hundred or so attendees arrived, they were welcomed with soaking wet tents, a lack of food, resources and facilities. In an instant, Fyre Festival had turned from a paradisiac getaway to a modern-day Lord of the Flies – with influencers desperately seeking a way out.

Similarly, and at the time I'm writing this, all three seasons of *The White Lotus* – a show that follows guests and staff at luxurious resorts, with at least one of them coming to a deadly end – take place on islands. In our minds, they hold the paradoxical feat of being at once places of utter relaxation, and of stranded nightmares. Season three of *The White Lotus* takes place in Koh Samui, another Thai island. The sprawling white beaches and tropical greenery, lavish cruise ships and zen monasteries, monkeys and the sound of chirping birds ... all of it contrasts with the first scene of the series, where gunshots are heard, staff are seen fleeing and cowering as a character stumbles upon a floating body. As a result, we as spectators are compelled to see all that follows in the series — the birds, the water, the monkeys and the plants — in a different, deathly light.

I hesitated over whether to talk about Koh Kra in my first novel. I still carried within me the shame of being a snorkelling tourist on a similar island, blissfully ignorant of the region's past. There was more, of course. The difficulties of writing about horrific events in a way that does not sensationalise them; the knowledge that my family would read the book, and that it would raise in them the same questions and affronts it had raised in me.

In the end, I did write about it. Not doing so felt like avoidance: as if I were trying to assuage the past, instead of facing it head on. And funnily enough, after reading the book, my mother asked if we could go back to Phuket. We were just coming out of the pandemic, and the five of us hadn't been away together for a while; so, we went.

I did not ask her why she wished to go back to Thailand. Perhaps it was a way to get closure. Perhaps she wanted all five of us to go back, now that we had fuller knowledge of the stories the region held, and see it in this new light. Or perhaps even after learning about Koh Kra, she still saw Phuket for what it had been, almost ten years ago, that summer when we had all gone as a family. A beautiful, calm island, with blue seas and friendly locals and soft sand under our feet; a place that held memories of joy and bliss. There is one thing, though, that she did not want to do while there: she did not want to go on any boats.

The rest of my first year in London went by quickly, in a cloud of anxieties and homesickness. But sometimes, in between all of it, I could just about glimpse moments of the joy and freedom I had so longed for. New friendships and romances, dancing until the early hours of the morning – the vertiginous realisation that I could eat only sweets or junk food for days on end, with no reprimand. Moments in which I felt unmoored from my previous self, free from family entanglements and heritage and expectations; moments in which I had the luxury to be adrift, while I tried to figure out who I was.

Twelve years later and London is still not exactly home, but close enough: my career, my partner, my closest friends, they all reside here, within the life I have built for myself; a life that emerged from that first year, one that existed in between desolation and delight, a year that quivered with prospects and a whole world opening up to me. And now, when I think of that year,

I think of it as holding some of my lowest days, alongside some of my happiest memories.

When I return to Madame Z's advice and try to picture my happy place, I do not picture one that is real. There might be the sea, yes, and the lulling of the waves and the warm sand – but I imagine it as a blank slate, hovering above the physical world: an island that exists solely in that hopeful state of becoming, ripe with utopic possibilities and the promise of idyllic days. The same state that I had been in all those years ago, advancing slowly in the blue sea, the fish gently nipping my ankles and my family waiting for me ashore, in between childhood and adulthood.

Jackfruit and Jaggery

SANTANU BHATTACHARYA

My parents live on an island. The sunshine is soft, the earth loamy, the winds rustle in cyclical eddies, giant palms sway in manic movements.

The fertile soil grants the place a rare abundance. People commune with plants, plants with the breeze and the blazing sun, the sun with the clouds, that then release their love in lashing rains. There is a flavour for every season: the cloying sweetness of mangoes in summer, the jackfruit that is juicy in the heat then turns meaty in the monsoons, the molten jaggery in the spring, the leafy greens in the winter.

This island is old. It didn't emerge from under the sea as a consequence of tectonic plates moving. Instead, it was put together by people who migrated from far and wide – kings and their counsel; warriors and their soldiers; priests and their scriptures; travellers and their quills; fisherfolk and their day's catch. They nourished and nurtured it, inch by inch, with contributions from as near as Mithila to the west and Burma to the east, and as far as Persia, China and Indonesia.

Once it came into its own, it carved a distinct identity out for itself. Patrons tilled the land, poets sowed seeds, writers sprinkled water, classicists drove the ploughs, musicians harvested the crop and singers picked the fruit. And the people turned up, on every occasion, to revel in the riches, to partake in the meals, to celebrate the successes with lamps and drums, *dholkortal* as they're called.

Soon, the neighbouring islands began acknowledging the greatness of this place, looking up to it, hailing the quality of its produce, the enterprise of its people. This island, for its part, never shut its doors on anyone. Everyone was welcome, the island better for it. Centuries later, this is where my parents were born. They've lived here all their lives.

I was born on the island too, living there for the first three years of my life. I might have spotted goods from other places, some familiar some foreign, but the ground we stood on was unmistakably ours. Even today, four decades later, this is where I meet my parents, no matter where we are in the world. On this island, I'm still a child, even though I'm now middle-aged; I'm imbibing the nostalgia of my mother's stories, the heft of my father's discourses, the passable humour of my cousins' jokes. Here, I'm comfortable; I'm not looking for new ways to be hot, desired, employable, modern. I can be lazy without being a failure, sarcastic without being insulting, respectful without being fawningly formal.

On this island, I'm home. This island is called: Bangla.

When I was younger, I thought that multilingual people lived in a vast ocean of words, dismissing allegiance to any one island; mermen and mermaids, water their natural habitat. But I'm now convinced that to be multilingual is to hop from island to island, laying claim to each parcel of earth with varying levels of entitlement, to be a different person in each, sometimes voluntarily, sometimes reflexively. This switch from one island to another could happen in mere moments: I might speak to an acquaintance in one language and immediately whisper a funny anecdote in my partner's ear in another. Which of those is me? Both, of course, for who else could they be?

English is my other island. I journeyed here when I was three, along with many other children; our parents put in a lot of work to get us here: queuing under the unforgiving tropical sun to collect forms at English-medium schools, preparing us for the school interview, pulling strings to find influential Board members or ex-students who could put in a good word. Some parents moved here, lock, stock and barrel, cutting out the distracting noises from anywhere else, creating 'English-only' homes, exiling their children from all other islands, including their mother tongue. There are stories of teachers hitting students for speaking anything but English; *uncouth*, *verni*, were just some of the words used for those who struggled with this foreign tongue, unable to find a foothold on this strange island. Thankfully, my parents, even in their wildest imagination, wouldn't think of abandoning Bangla at home. And so I learnt to gracefully flit between these two worlds, English by day, Bangla in the evenings and weekends.

On this island, I am a completely different person to who I am in Bangla. I am global and smart; I know the latest lingo, at the workplace, on the street, at social engagements and with a lover in bed. Trained meticulously by my teacher in the Queen's English, I used to be a thorough student, picking out bombastic vocabulary to justify my claim to this foreign soil, as so many of us from post-colonial societies do. But the adult me, as he unpeels his layers of reserve to embrace his sexuality and creativity, is also learning to express himself in ways that are sassy, saucy, satirical.

Since my move to Britain ten years ago, I've had to reconcile with living a largely monolingual life. It felt uncomfortable at first, to be hemmed in by these barbed-wire borders. I was so used to island hopping, choosing words from here and there to give my sentences the best chance, to breathe essence into them. Here, I'm bound by the flora and fauna of English, its weather and ways. The natives rarely speak another language. If they do, they're reticent and self-conscious. When they do, they're celebrated for it.

Sometimes though, in the middle of a nondescript day, I catch a whisper of other islands – an old schoolfriend prattling on in Bangla, someone speaking Polish on the phone, two aunties conversing in Swahili on the bus, a group of Turkish youngsters returning from a night out – and I'm reminded of how cross-pollinated English is, furrowed and fertilised by birds and bees, the oceans blowing in seeds and fruit from faraway places, everyone moulding the damp earth into shapes that reflect their cultures, their histories. In moments like these, I don't strain myself to find the right words, to get the pronunciations right, to worry about how I'm coming across to a local; instead, I stop second-guessing myself – the immigrant's plight – and find the confidence to say, 'I'm not articulating this well enough in English', something I'd never admit when I first came to Britain. Sometimes, all it takes to feel at home is to be at peace with, perhaps even flaunt, one's shortcomings. And it can take years to get there!

It won't come as news that English has a troubled relationship with Bangla. The two islands I inhabit most often are separated by waters choppy from decades of tumultuous storms. So much so that Bangla has even been robbed of its original name! The colonialists called our land 'Bengal' and our language 'Bengali'; in our tongue, the word for both is just 'Bangla'. For centuries, we've been living a double identity – one for ourselves, one for the world.

The Partition of Bengal in 1905 by the British colonial government was not just a territorial division (into its eastern and western halves) or a religious one (Hindus went west, Muslims went east); it was also a ripping apart of our lingual fabric. No island is homogenous – one part might be dry and shrubby while another under the cover of lush vegetation; likewise, the Bangla spoken in each of these parts has distinct dialects and accents. In the east, the language has more 'z' sounds, is more relaxed to the ear, and includes a smattering of Arabic words given its Muslim population (Bengal was a key port on the trade route between the Middle East and

Southeast Asia). The Bangla in the western half is more clipped, nasal, consisting of stretched vowels and missed consonants. These two dialects, enmeshed for generations, were sequestered in their respective halves by the Partition of 1905, and further formalised into international borders in 1947 when India and Pakistan were born as independent nations. The border was drawn at the whimsy of a British mapmaker who dispensed of this mammoth responsibility from his distant dwellings in New Delhi, without once visiting the lands and people he'd segregate for eternity.

My father's family are from *o-paar Bangla* (the part of Bengal that falls outside India). That tract of land, though topologically unchanged in its alluvial soil, wide rivers, fertile delta and dense mangrove, has undergone a series of political metamorphoses in just over half a century – first named East Bengal in 1905, then East Pakistan in 1947, and finally Bangladesh in 1972, when it became its own nation. Its brutal fight for independence, first against British colonialists and then against Pakistani theocrats, has always been grounded in its innate need to give its language pride of place. No surprise then that it is one of those rare countries that draws its name from its mother tongue: Bangladesh – the home of the Bangla language.

My grandmother arrived in Kolkata, a young girl, from what was then East Bengal, accompanying her chronically sick father, mentally unstable mother and seven siblings, all heaving from the weight of the land, home and language they'd left behind. While the currents of the 1905 Partition caused them to migrate, the 1947 Partition cemented their displacement. Decades later, when my father, at the ripe age of fifty, crossed the border from India into Bangladesh for work, he finally reversed the winds that had carried his ancestors in the opposite direction. His client, upon learning his origins, excitedly drove him from Dhaka to Brahmanbaria in Comilla district where our family hailed from. My father called his mother from the village to tell her where he was. My grandma, overcome by emotion, asked my dad to pay his respects to the land by scooping up some soil and smearing it on his forehead. Then, she asked him to say something in the local dialect. 'To whom?' Dad asked, 'I don't know anyone here!' To which grandma replied, 'Say it in the wind.'

Back then, as a teenager, I'd seen this exchange as sentimental. But now, I know the sharp stab one feels when walking out on the street day after day and never hearing the language always on the tip of one's tongue. My attempt to write stories in English, tales of my land and my people, to give them a new lease of life, is therefore my own little act of defiance. I have even begun taking on translation projects from Bangla to English, planting the riches of one island in another's soil, then standing back and marvelling at what grows. It is a manifestation of my grandma's words, a sort of poetic justice – say it in the wind, so it may travel far and wide, so it may never go unsaid, so it may not die with us.

Sometimes I visit other islands. These sojourns are elective, some more organic than others. Hindi is a staple since childhood. Spoken widely across northern India, I picked up the language on the streets, in conversations with shopkeepers and taxi drivers, but mostly during the endless hours I spent watching Bollywood films as a starry-eyed kid. No surprise then that on this island, I am not a real person but a fictional character; peppered with a generous drizzle of the lyrical Urdu, I say my lines: a little romantic, indignant, melodramatic, laced with a flourish that would be difficult to find in everyday conversation. It's like starring in my own Bollywood film! In French, an island I've visited for short periods as an adult, having periodically enrolled for lessons, I am a perpetual outsider, forever a lingual immigrant, whispery and demure, but also courageous in ways only an immigrant needs to be, stringing words together, every sentence an achievement.

Each island gifts me unique riches that I'd be hard put to find anywhere else. In Bangla, the funny twin adjectives *nyaka* and *paaka* respectively describe the traits of being benignly over-dramatic and being precocious like an over-ripe fruit. In Hindi, the noun *thairaav* signifies the state of standing still while the world moves in swirls around us. In French, the verb *aimer* is to like and to love, a practical reminder, in ways that only the French can, that the most celebrated sentiment better come with a pinch of salt. In English, I'm obsessed with how alive and young the language remains: words that didn't exist a generation ago are constantly manifesting out of thin air, and words that did exist are being remade for the future to claim in new ways.

There is a lot on the internet nowadays about multilingualism, posts and podcasts propounding its many virtues: it makes us 'empathetic', it helps us 'understand cultures', children's minds 'work better', adults can 'delay onset of dementia'. Maybe this is all true, but I resist the need to intellectualise and 'make a business case' for something this fundamental to our existence: language differentiates us humans from other animals on earth. A burgeoning vocabulary bound by the orchestration of grammar, the flexing of sounds to accentuate and diminish emotions, and to do it over and over again, across generations, in regions and corners around the world, through accents and dialects and proverbs and jokes – we don't need memes and reports to tell us that language is possibly the greatest gift of human civilisation to itself. To be inattentive to language, to not pass this treasure trove down to the next generation, or worse, to blatantly dismiss it as the leisurely pursuit of an elite class, as I see happening increasingly, is to dismiss what is at the very core of who, where, why and how we are.

With the passing years, my sister and I have trained our non-Bangla-speaking partners in elementary Bangla, relieved that they've picked it up well, speaking more and more of it in our homes. I've glimpsed it in my parents too, that clichéd need to go back to their roots as they grow older – geographical and cultural yes, but also lingual. With age, they're holding on to Bangla like a log in a flood, as though their island is now besieged by water. In fact, Bangladesh is now at the forefront of the climate crisis, its low coastal land at risk of being submerged by the rising seas. If indeed one day the land ceases to exist, what will remain? Will it not be us and those after us, building islands of Bangla in our homes, keeping alive the memories and heritage of our forebears whose blood once irrigated the land in service of its beloved language?

On a recent visit to the doctor, my mother described her condition completely in Bangla, knowing that the doctor didn't understand the language, even though she herself gets by just fine in English. As I translated her monologue to the medic, slightly irritated at first, her refusal to describe her suffering in a foreign language struck me as most natural: her subconscious mind could reach into only one lexicon to share her deepest fears of decline and death. It was scary, yet beautiful. At that age, maybe there is no need to mourn the loss of linguistic flexibility; we have already been so many people, lived so many lives; those of us who have hopped from one lingual island to another, getting on small boats, inflatable dinghies, crammed airplanes, still carry the tastes of jackfruit and jaggery on our tongues.

At the end of every story comes the time to journey back home, and to bow our heads to the stuff from which we were created.

Leaving Avalon

RALF WEBB

I grew up a little under an hour's drive away from Glastonbury, a small town in Somerset, that famous West Country county of dwindling cider orchards and Ministry of Defence bases. I visited many times as a child and teenager, often with my dad. We'd wander around the high street, drop by the esoteric booksellers, crystal vendors and health food shops, and climb the hill that rises above the town. Approaching Glastonbury from the north, along the Wells Road, this hill can be seen in the distance; the eye snags on the stone tower at its summit. Rising from an otherwise flat landscape, one's first instinct is that the hill, with its smooth, whale-backed slope and terraced ridges, must be man-made. If it is dawn (and the weather is just right) a mist hangs above the surrounding fields, and the hill, perforating this mist, seems to hover in the sky, a mirage or visual glitch that severs its connection to the earth and reality entirely.

The hill, which climbs some five hundred feet above sea level, is known as the Glastonbury Tor (*tor* meaning 'high rock', likely cognate with either the Gaelic *tòrr* or the Old Welsh *twrr*). The tower at its summit, St. Michael's, is the remnant of a fourteenth-century church which was destroyed in the aftermath of the Dissolution of 1539 (its abbot was subsequently hanged, drawn and quartered inside). The Tor is situated on the Somerset Levels, a vast low-lying coastal plain which once consisted of peat bogs, marshes and wetland, veined with rivers and streams. In the Middle Ages, serious land reclamation and drainage efforts were undertaken by local monasteries in order to make it suitable for agricultural use. But the Levels remained prone to flooding; a problem that persists today, despite modern efforts to mitigate it. Moving into our new, chaotic era of escalated precipitation, severe flooding has become more frequent. During such deluges, the Tor occasionally shifts into the past, resembles something that it had once been: an island.

An astonishing, contradictory medley of myth and superstition has been heaped upon this land-locked island over the millennia. One of the Tor's foundational myths links it with Joseph of Arimathea and the Holy Grail. Another suggests that the Tor is the Isle of Avalon, a location in Arthurian legend where the sword Excalibur was forged and the mystical enchantress, Morgan le Fay, was said to reside. Many of these Arthurian standards were created and perpetuated by Geoffrey of Monmouth, in his twelfth century *Historia regum Britanniae*, an influential work of literature that melded history, fiction and prophecy to spin a fantastical saga of Britain's lineage. The *Historia* connected Britain with Troy, portrayed the Britons as triumphant conquerors, and promoted the progress of the Christian faith. Like Virgil's *Aeneid*, Monmouth's *Historia* imparts racial patriotism and fortified national identity by elevating the idea of 'Britain' to that unassailable realm of myth and divinity. But the Tor was not linked to the Isle of Avalon in the *Historia*; the monks of Glastonbury Abbey made that connection, claiming to have uncovered Arthur's grave at the hill's summit. This was, of course, fake news: a pseudo-archeological discovery which aimed to attract wealthy pilgrims to the monastery, a myth dug up from the soil for financial gain.

Countless other decidedly far-out myths and superstitions arose around Glastonbury Tor throughout the late nineteenth and twentieth centuries, carried by various waves of Western Esotericism, and continue to attract pilgrims and spiritually-minded visitors to this day. Among them, that the Tor is hollow and contains a dragon, crystal cavern or gateway to the afterlife; that the Tor is part of a terrestrial zodiac; that the Tor is a node on a Ley line and thereby conducts or channels 'earth energies'. Whatever one's opinion of such ideas, it can safely be said that this island is a site of enchanted thinking. The island is a fiction, or a haven for fictions. The stories and legends attributed to it bloom, flower and then putrefy; they leech into the earth and alter the environment. The resulting locale is neither real nor unreal, but an in-between place; an island that hovers in the imagination.

My dad's interest in Glastonbury's latter-day lore was only superficial – he didn't drink the New Age Kool Aid, so to speak – but his attraction to the Tor's Arthurian associations was sincere. On our visits, he would namecheck King Arthur, the Knights of the Round Table, Merlin, and so on, always in a deep, self-mocking tone of voice that made light of these references while somehow conveying a degree of high seriousness, like a knowingly bad parody of Christopher Lee. I'm not sure what it was, exactly, about Arthurian legend that fascinated him; maybe only that it promised to tincture the brute, violent reality of Britain's Christian national identity and heritage with magic; to transform a land of gloomy Protestantism into a dreamscape. Such a transformation would no doubt appeal to a man who was an unequivocal atheist ('Don't worry,' he said to me, in one of our last conversations, when he knew that he would soon die, 'I'm not turning religious, or anything.'), and yet who also felt a strong attachment to his identity as a 'country boy', which is to say, a countryman.

Glastonbury's unavoidably alternative social fabric also resonated with another side of my dad – his engrained anti-authoritarianism and evergreen hatred of the right wing, which was, I remember, almost spiritual in its fervour. In his youth, he was a rocker-turned-punk-turned-post-punk acolyte; he wore a pin in his ear and drainpipe jeans 'before they were cool'. The rebelliousness of spirit that I occasionally saw in him, growing up, must have only been a relic of what it once was, perhaps diminished by the responsibilities of adult life, the disappointments and crimes of New Labour, and merged – uncomfortably and incongruously – with his career as a secondary school teacher. Mostly, it found an outlet in music, which he liked to listen to very loudly while driving very fast. On those drives to Glastonbury, he'd play the same CDs on repeat, drumming the steering wheel: Caravan, Steely Dan, Echo and The Bunnymen, and Van Morrison; the cracked CD case of *Avalon Sunset*, with its cryptic cover art of a swan on a crepuscular lake, left a weird imprint on my childhood memory.

But if he was pulled to the Tor because it spoke to or rekindled this youthful nonconformism, this aspect of the place also seemed to repel him. His relationship to the town (or his appraisal of its pilgrims) was skewed. The tie-dyed, barefooted, wayfaring habitués of Glastonbury would frequently cause him to widen his eyes at me, or to arch an eyebrow; private, collusive signals that meant to convey prejudice and astonishment, that meant to say that he and I were different from them; that we were better than them. My dad had aspirations. He aspired to a certain lifestyle, a certain level of comfort. If British social class can be thought of as series of territories with volatile borders, I'd place my family on the very outer edges of the middle class: my dad wanted to migrate to the interior. I think that his frustration at being unable to do so clouded his kinder instincts, hardened into censoriousness. He was so quick to look down on others, on strangers, on anyone, a quality that was aspirational in itself, because it mimicked the behaviour of those in the interior: the very British, passive-aggressive incivility of the well-to-do, people who believe that the denigration and judgement of others is a canny investment, or a surety, the only thing mooring them in affluence.

LEAVING AVALON

My memories of those trips with him are impressionistic. I can only recall brief images – thrifting a copy of Ted Hughes' *Crow* in one of the bookshops; neon-green slime in the stone watering trough in a field at the base of the hill; everything, in fact, very green and lush. I was seventeen the last time he and I climbed the Tor together. It was a little under two years before he was diagnosed with a rare form of leukaemia, an aggressive and terrifyingly efficient illness which killed him just nine months later. I wonder if it was in him even then, whether those mutations were already beginning to accrue inside his marrow. With typical teenage solipsism, it did not occur to me that these excursions to the Tor were significant to him, and I was often moody, uncommunicative, harbouring private ideas and opinions that could not possibly (so I believed) be understood by others. Or maybe I was irked by his aerobic fitness, the unblinking determination with which he climbed the Tor along its steepest gradient. As an athletically disinclined teenager (especially one who also wanted to be a poet) I would have rather wandered along at a dreamy tempo, or loafed in the grass, chasing castles in the air. My dad's quintessential trait was meeting the outside world head-on. He seemed to take pleasure from punitive feats of outdoor endurance (daily long distance running in all weather) and astonishing stupidity (hiking in heat-waves without bringing water, because 'it won't take long'). Traits that, as an adult, I have come to acquire, too, and which astrology-minded friends have deemed 'Aries behaviour'. I would ignore this astrological evaluation entirely, were it not for the cosmic coincidence that I was born on my dad's birthday, and feel in my bones – or wish to believe – that this set my attitude to the world at a similar angle to his.

I found a photo of him, recently, standing near the top of Glastonbury Tor in amber light, holding me as a newborn. The lichen-specked St Michael's Tower rises behind him. He's wearing Wayfarers, a sweatshirt, jogging bottoms and boat shoes, the ensemble of a new father who can't quite let go of the 1980s; who hasn't quite let go of his youth. Sheep are dotted around him on the green slopes, like balls of cotton wool. In the background, where hill meets sky, there's a curious, out of focus figure: a person dressed in white, evocative of a ghost or angel, whose indifferent incursion on this scene makes the photograph, somehow, seem so alive, rather than a too-perfect, uncannily cool portrait of a man who is no longer here. The figure in white arrests my attention – its ethereality lends a quickness to the image, turning it from a static relic into something stranger, something that still shimmers with a wonderful and inexplicable light.

The photograph also testifies to the presence of its photographer: in this image, the shadow of the photographer, my mother, falls upwards across the grass, inclining towards me and my dad. When I asked her about this photo, hoping for more context, she didn't really remember taking it. She even suggested that the baby, half-hidden behind a sling and wrapped in several layers, might be my sister rather than me. Worried that I was disappointed by this, she quickly recanted, feigned a moment of revelation, and declared that it was me after all, she was sure. But of course she wasn't. The photograph has become unfastened from the moment in time which it captured. It has become unattached from the memory of its photographer, and the memories of the subjects it depicts. It is an incomplete or partial document, an incomplete or partial history, and in its gaps and absences I am free to invent and cultivate my own story about its origins and imbue it with as much or as little significance as I wish.

I went back to the Tor last summer for the first time in sixteen years. The double decker from Bristol, so like a passenger ferry with its lumbering bulk, bobbed and jostled on the ninety-minute journey. The road curved past Apple Tree Glamping; a billboard advertising the

Lancelot Estate Agent sale-by-auction of 250 acres; a bungalow whose front porch was festooned with stone statuettes of dragons, buddhas, caricatures of Confucius. The roadside trees were in verdure (oaks, elms and willow, many choked by ivy) and as the bus neared Glastonbury, the Tor and its tower flickered in and out of sight, disappearing and reappearing through curtains of leaves.

A quick roam around the high street suggested that not all that much had changed. Only, I sensed a latent edginess, the kind that swells in humid English summers just before a storm breaks. Everything felt slightly peculiar. A vintage shop was offering fifty per cent off its fifty per cent off sale. At the bottom of the high street, a dozen or so people, decorated in spirals and streaks of body paint, danced as a white guy played Bob Marley on a fuzzy, electro-acoustic guitar. Beside them, there was an enormous inflatable cube, its PVC material patterned with the iconic strings of neon 'code' seen in *The Matrix* films. On each face of this cube, iPads had been crudely attached, their screens showing an icon of a red pill. Occasionally, a passerby approached the cube and tapped the pill on the screen. The guitarist played 'Get Up, Stand Up'. I headed for the Tor.

At the base of the Tor are two natural sacred springs, both of which are said to have magical properties. A

Victorian well house covers one of these springs and has been transformed into a place of spiritual worship. Realising that I had never actually been inside on any of my prior visits, I stopped in en route to the hill. In the dank, dripping interior, I found a group of women in white gowns, garlanded with flowers, bathing and chanting in the central pool. Thick white candles had been placed around the edge of this stone basin, some so knotted with bulbs of melted, hardened wax as to resemble ghostly tree stumps. Knowing nothing about this ceremony, I had no particular opinion on it, but I felt my presence there was its own statement: I was a spectator, which made these sermonising, bathing women a spectacle. It was hot by the sacred spring. The air felt extremely close. I watched for another minute, and then exited.

I climbed the Tor along the steep gradient. At the summit, I lay down in the grass, ate an apple, and looked out across the Levels. The greyish, overcast sky and green fields merged into one another. Sheep drifted across a pasture at least a mile distant, like surreal tadpoles in a shallow pond. There were a number of tourists at the top of the Tor, picnicking and taking photographs of the views and tower. I overheard snatches of Chinese and French, and the frontiering registers of American-English. A middle-aged man in beads and outsize hemp clothing spoke at two young, fairycore women about universal consciousness and Gaia and so on. As the women made moves to leave, he insisted on an embrace. They capitulated, and he held on several beats too long, utilising the innocuous, impotent guise of male spirituality as an expedient to cop a feel, like a cult leader will leverage power through misogyny by first labelling it 'love'. The women hurried away. The rain broke. I pitched the apple core over the edge and left.

The descent from the Tor – if one descends along the gentler incline of its southern slope – is inlaid with large stone steps. As I walked down them in the drizzle, I felt a twinge in my knee, and had to stop for a moment to stretch. And then I experienced a very clear memory. I remembered, on our visits, how my dad would always complain that these steps were awkwardly spaced; too large for one stride and yet not large enough to comfortably accommodate two, such that descending them forced an odd canter that put a strain on his knees. These complaints, I recalled, seemed to grow in intensity with each visit, as if every trip to the Tor was a lesson in the corrosiveness of time – its attritional quality – how it wears out and fatigues the body. My adolescence, I remembered, felt so marshy and vast – it was something I had to crawl out from or else perish in – and yet my dad, traipsing down this side of the island, made adult life seem intensely pressurised, something brittle and liable to break, something that rushed towards an end. Recalling this, I was struck with another realisation; I am the age he was, in that photograph, when he stood at the summit with his child in his arms. And I had the eerie impression that as I lurched down the stone steps I was passing in front of him, I was passing in front of a point in his life, like two planets in transit, and that when I reached the base of the island I would no longer recognise where I was, but would be so happy that I had come.

Take Me Oh Sea!

ELLA FREARS

If you were to travel to St Ives in Cornwall by train — a small train that hugs the coastline for ten minutes from St Erth — you'd find yourself first gliding along Hayle Estuary with gulls and waders picking through the patchy clumps of grass, before sliding west onto windswept sand dunes, past dog walkers, families, people out for a run; you would see the endless strip of Gwithian beach stretching out behind you and Godrevy Lighthouse at its furthest end on a rock just out to sea; you would curve in and out of the pale yellow beaches of Porthkidney and Carbis Bay, catching glimpses of St Ives harbour ahead, the view dipping behind trees, disappearing as you pass through a tunnel; you would slide around the clifftop and through a stone railway arch and there you'd see St Ives startlingly in full view — the many wide-eyed cottages turned towards you as though crammed in and smiling for a photo; finally, you would pull into the station above Porthminster beach with its small, green strip of grass and scrappy palm trees leading down to golden sand and, on a good day, a perfect turquoise sea. If you arrived this way — the sea, the sand, the sky, the harbour curving round to greet you — you might think that this was one of the most beautiful places you'd ever seen. And you'd be right — it is beautiful, objectively so. Though, growing up in St Ives — a restless teen who dreamed of the city — I almost took pride in my failure to be awe-struck. It's taken fifteen years of living in London to shrug off my forced resistance to my hometown's charms and accept the many moments I was, in fact, swept up in the majesty of it.

Everyone has a personal map of their hometown. Private topography featuring memories, happenings, emotional events. I could take you back through that train journey again and instead of sand dunes, gulls and cottages, I could point to a school friend's house, an illicit but anticlimactic rave, a disastrous skinny-dip, a dentist who pulled nine of my baby teeth in one go, a near-death experience with an antisocial llama called Augustus, a Christmas Eve kiss under a streetlamp, and countless sites of hanging around and being moved on by adults — what else was there to do? Among these locations, there is one place that feels like the epicentre of those memories. A place where all aimless wandering seemed to lead: The Island.

The Island is not an island at all, but a promontory — a small hill with steep cliffs along three sides and a little stone chapel, St Nicholas, at the top. The chapel dates back to 1437, though it's thought that there was a structure there as early as the fifth century used by Irish missionaries. You can find The Island in almost every famous photo of St Ives — a bare, green shape, connected by a wide strip of land, at odds with the crowded houses, cobbled streets and beaches that make up the rest of the town. Setting off on a walk around St Ives without agenda you'll inevitably end up there. The path around it is short. You could loop it in five minutes, though I'd recommend dawdling, looking for seals, or staring out to sea with narrowed eyes like one of the customs men who'd watch for smugglers from the chapel walls in the eighteenth and nineteenth centuries. No longer a covert port for illicit goods — unless you count my halcyon days of poorly rolled joints and bottles of Smirnoff Ice — The Island is less a destination in and of itself and more of a track — a detour. Pulling into St Ives on that train, The Island is visible peeking over the houses and harbour. It does something to the scale of that view, narrows the distance, flattens it and makes the houses seem novelty-small; the boats tied in the harbour, either beached haphazardly or bobbing jauntily, like toys.

When I think of this view and its strange scale I think of the painter, Alfred Wallis, who lived and worked in St Ives in the early twentieth century. A former seaman who started painting at the age of seventy after his wife's death, Wallis conjured the sea and surrounding landscape in a style that has been described as naive – though that doesn't capture the clarity of vision, originality and charm of his work. In his painting *The Hold House Port Mear Square Island Port Mear Beach*, The Island features as a dense but fuzzy green shape with exaggerated curves and black rocks around its edges. It's a striking form against the grey, black and white of the rest of the image. The size of the houses varies according to importance as is often the case in his work, and Wallis is less concerned with the accuracy of the view (you can see around a corner in this painting) than with the feel of it. I recognise the crush of beach, cottage and sand alongside the weirdness of that grassy hill jutting out to sea; the way certain locations on your personal map loom larger depending on the emotional charge they hold.

I've never walked around The Island anticlockwise. I don't know that there is any particular reason for this, just that it would feel perverse for me to do so.

I always approach via the stone slope leading up from Porthmeor Beach, another pale stretch of sand flanked by The Island on one side and the rugged cliffs of Clodgy Point on the other. Arriving from this side the path runs parallel with Clodgy Point before bending round to bring you face to face with the sea and nothing else. This is where the ocean feels at its wildest, its largest. It's also where you can look for seals which you will often find looking right back at you – their little bald heads, shiny eyes, and dark snouts, bobbing inquisitively. The rocks around The Island are mottled grey, white and black and some have patches of bright orange lichen – the same you might see on the roofs in St Ives. The path then slopes upwards towards the Coast Guards' lookout and round next to a granite battery built in the 1850s to protect from French invasion. Here, you could leave the path and walk up the grassy bank to the chapel, or follow it down to the car park above Porthgwidden Beach (smaller than the previous beaches mentioned but still lovely). This car park covers most of the land connecting The Island to the town. Parking is a contentious topic in St Ives, there being very little of it. Many St Ives car parks are notorious for clamping the cars of families who only overstayed by a few minutes. I remember a local news story in the early 2000s of a farmer depositing a bale of silage in a parking space with *CLAMP THIS* painted on the side. At the bottom of the car park are a few buildings that are technically also part of The Island. One of these is a low oblong building – St Ives Judo Club – where I spent most Monday evenings between the ages of six and eighteen.

St Ives Judo Club was founded by Major Mike O'Neil in 1963 who had been introduced to Judo while on a boat returning from the war in Korea. When I joined the club aged six he was still teaching, though he seemed like the oldest man in the world. I remember him leaning down with some effort and saying, 'If all else fails, buttercup, just kick him in the humpty-doodles.' The club was sparse and a little run down. Green and red squares of mats on the floor making up a large rectangle. The walls were bare, red brick except for a thick carpet, somewhere between oatmeal and grey, which had been hung halfway down them to the floor to create a soft barrier in case you were sent flying. Kneeling at the edge to watch the sensei I would lean back against the fraying carpet and tug at its fibrous edges. When I think of judo, it's the textures that I remember – the dry sweep of the mat underfoot, the coarse walls, the feeling of an opponent's heavy cotton sleeve tugged from my grip. In judo, one of the first things you're taught is how to fall. I didn't like this. I dreaded being thrown. I was spectacularly winded by the same slightly older girl week after week. I liked the groundwork though and would take any excuse to pull an opponent to the floor so that we were wrestling there instead. I loved the feeling of holding someone down, sensing their movements and shifting my weight accordingly. I was never particularly gifted, but I liked the physicality of it. It is an intimacy that is different from any other I've felt since. Non-romantic, non-sexual, rules-based but still thrilling in a bodily way; defined by closeness. Judo translates as 'the gentle way' from Japanese – a fact Major Mike O'Neil liked to remind us of, if we were getting scrappy.

As a child I was boyish. One sunny evening, walking alone through Trewyn Gardens – a small, gated park – to the judo club, wearing my gi, I found myself suddenly surrounded by older boys. 'Look at the Karate Kid,' one of them said, 'come on, show us some moves.' They were closing in and I realised they thought that I was a boy and that they were going to hit me. It seemed more embarrassing to correct them than to be beaten up. Just before they grabbed me, I somehow managed to duck and sprint away. I never wore the gi outside of the club again. I also started making a conscious effort to look like a girl – cue a whole host of other issues, though the judo did come in handy for some of that ('just kick him in the humpty-doodles'). The St Ives judo club logo depicts one judoka throwing another in front of a green shape. I've always thought that this shape was The Island, though looking closely now I realise that it's a leaf. Still, that's how I picture that judo club — a small low box under the quiet hill of The Island.

My parents recently discovered the teenage diary of my paternal grandmother who died when I was eight. It's a small black notebook with a red cloth spine onto which she's inked *Diary 1945. WAR. DEATH.* She was sixteen, from Manchester. On 22 August she wrote about a trip to St Ives with her friend Esther: *I never imagined there was such a peaceful place on earth.* This was a revelation. My mum and dad, from Manchester and Leicester respectively, met as students in Sunderland and after travelling together, settled in St Ives because of my mum's fond memories of childhood holidays there. My dad didn't have a prior connection to St Ives and as far as he knew, his parents had never even visited before. In this diary, however, my grandmother describes a transformative walk around The Island at night.

> *I felt as if I were really living. I felt insignificant and powerless against the elements . . . Ah! I shall never forget that night. I scrambled over the rocks until the spray splashed against my face and warmed my blood. The noise was terrific I felt as though I wanted to infuse my body into the hurling mass,*

TAKE ME OH SEA!

I tried to shout above the roar of it, 'take me oh sea!' but the sea itself drowned my excited cries.

Her handwriting is wild and looping on these pages. I have very few memories of her – mostly they are of visiting her towards the end of her life, ill in bed, struggling to breathe. It was thrilling then to find her there in the diary – sixteen, on a stormy night at the top of The Island, alight with the beauty of the place. Three months earlier she had stepped out of her family home in Manchester for VE Day – *oh boy at last hostilities have stopped in Europe! People were laughing and laughing.* She was on the cusp of womanhood, emerging from six years of war. It must have felt like anything was possible. The chapel where she stood is named St Nicholas after the patron saint of sailors. In less than ten years she would be married to a stubborn but kind sailor from the Navy. Whether she knew it or not, St Ives in 1945 was exciting – international artists such as Ben Nicholson, Barbara Hepworth and Christopher Wood had settled there during the war (Nicholson and Wood 'discovered' Alfred Wallis there). For that period St Ives was listed alongside Paris, New York and London as a cultural hub. In part this was to do with the magic of the place – the blue quality of the light that is specific to St Ives due to its being surrounded by sea on all sides. I like to imagine my teenage grandmother looking over the same landscape that inspired those artworks, perhaps even passing Russian constructivist, Naum Gabo on her walk through town.

In August 2008, exactly sixty-three years after my grandmother wandered The Island at night, I was dumped via text. I was sixteen too and had just returned from a family holiday in France. It wasn't a great love, and I would probably have felt OK about it, except that I'd sent this boy a postcard on the last day of the holiday. Now, looking at his breakup text, I was stung not by heartbreak but by my own words, my chatty tone that would arrive in a day or so. I winced remembering the lines I'd written about the Matisse Museum and how I thought he'd like it; the final 'I've really missed you!' My stupid exclamation mark. Restless and tearful, I left the house and headed, not to The Island, but to Clodgy Point. It was stormy and the evening was darkening. I climbed to the edge of the rocks looking out at the tumultuous sea and I began to cry, though quickly realised it was too wet and windy to cry properly. I stared out across the beach to the dark shape of The Island, the silhouette of the little chapel on top, and felt suddenly and deliciously insignificant. I didn't keep a regular diary, I never have, but I've always written in notebooks. In the little black Moleskin from that year I've written, *the wind whisked my tears away and out to sea*. It's bizarrely close to my grandmother's, *the sea itself drowned my excited cries.* In a film, you might overlay our timelines, show us – two teenage girls facing one another across the stormy sea – struck simultaneously by a bolt of insignificance in the face of the sublime.

Like her I was on the cusp of womanhood. Not emerging from war but unknowingly in the midst of a financial crisis that would eventually lead to the degradation of Cornwall's public services – death by a thousand cuts. Still, in a year I would fall in love with poetry via Sylvia Plath's 'Daddy'. In four years I would fall in love with a builder from Falmouth, sending him many postcards, letters and messages, eventually convincing him to move to London with me. The St Ives of my adolescence wasn't exactly the cultural hub that it had been during the war, though there was and still is a rich art scene. The Modernists and their legacy were ever present, just not always as reverently observed as they might have hoped. As a child I remember watching drunk men attempting to wiggle through the narrow aperture of the Hepworth sculpture outside the town hall. St Ives, like any town, was always changing. But The Island that I knew, that I know, is the same one that the Irish missionaries, the smugglers, the customs men and soldiers, and Nicholson and Wallis and my grandmother all walked around.

It would be different, I think, if it was called something else – The Hill, The Green, The Mound. There's something mythic about going to *The Island*. The name makes it sound separate from the town, the mainland – to be drinking on The Island, wrestling on The Island, watching for dolphins from The Island carries a specific romance that has everything to do with scale and insignificance. When you think of an island you imagine a small spot of land in the middle of a great sea. Standing at the far end of The Island and looking out at the horizon, you can imagine, for a moment, that what you see is all that there is. You are a tiny figure, alone, facing the elements like Miranda in *The Tempest*. Nothing else has to matter because those things are not on The Island with you. It's a compelling feeling – one even the most reluctant teen could get caught up in. One that might steal your tears or impel you to shout *take me oh sea!*

At the end of her entry for 22 August 1945, my grandmother writes: *These moments of supreme happiness that one feels in one's life last forever. Even now whilst writing this I feel excited, so I must stop now or I shall never sleep tonight.*

Island as Home

As in a Sea Parenthesis

SINÉAD GLEESON

At the end of 2024 I found myself, unexpectedly, writing about Napolean. Specifically, about his death. The cause of which has been much disputed, but I've always been intrigued by one of the more fanciful theories. After losing The Battle of Waterloo in 1815, Britain exiled the former monarch to the island of St Helena in the South Atlantic. Under house arrest in a forty-room wooden home, he had it decorated in his favourite colour, green. In the early nineteenth century, the main ingredient in most green paint, wallpaper and leather was Schweinfurt green, made from copper dissolved in arsenic. The humidity on his island prison led to the release of fumes, and when Napolean died at fifty-one, there were reports he'd been poisoned by his own wallpaper. Arsenic samples were found in his hair – not enough to kill him – and the likely cause of death was stomach cancer.

The debunked story of this murderous green lingered and I found myself thinking a lot about St Helena. Considered one of the world's most remote islands, it lies 1,200 miles from Africa, 1,800 miles from South America and measures just ten by five miles. Discovered in 1502 by the Portuguese, it was a stopover spot, a transitional place for ships and sailors before the Suez Canal was built. Its airport only opened in 2017; before that, it took six days by ship to get there. Communication is powered by satellite, and in our hyper-connected, less geographically remote lives, it's hard to believe that submarine fibre cable for broadband only arrived in 2023. Its most recent census lists the population at just under 4,500.

Hard-to-reach, sparsely populated locations hold a fascination for many of us. We idealise cut-off places, but what was it like for St Helena to become a society when the Portuguese landed, after centuries of being uninhabited. What does that do to an island? (Hint: the introduction of invasive species alongside deforestation contributed to the extinction of numerous native plants and trees).

Last year, I published a novel set on a small isle. Despite the numerous Irish islands I've visited which have influenced my work, this unnamed place does not exist, except in the pages of *Hagstone*. Its coordinates are never given, nor is there an explicit admission exactly that it's even *in* Ireland, but many readers assumed it to be a real place. There are around 250 islands off the Irish Coast – depending on your definition of an island – from tiny rocks to populated archipelagos. The number of inhabited islands with no road or bridge connection is thirty. The largest and most populous, Achill, is technically no longer an island, thanks to a bridge built in 1887, somewhat undermining its islandness. It lies off the coast of Mayo, north of Clew Bay, once said to have been home to 365 islands, one for each day of the year. This story from mythology made me feel justified in making up an island from scratch. But it's more accurate to say that my island is not *completely* fictional. Its tides are like those of Clare Island; its beaches borrow from the Aran Islands; its one-street strip is strikingly similar to Bere Island (a place I only visited post-publication for a book festival).

This fictional setting is the fulcrum of the book. It existed before any of the characters or the story, and in my head alone. The protagonist is a land and performance artist, and her life eventually intertwines with a commune of women living in isolation on the island. It took a very long time to write. It wasn't written consistently and was frequently interrupted by work, life and other books. But whenever the novel seemed to get swallowed up by distractions, I would just think of this island – sea-sprayed, meteorologically tumultuous, isolated – waiting for me to tell its story.

It would have been easier to base it on a real place, to dispense with the effort of invention. Somewhere like Donegal's Arranmore with its lighthouse and impressive cove, reached by steep, stone steps. A flimsy rope strung along rusted poles is a very casual approach to safety. The small bay was used for landing oil for the lighthouse, but could pass for a smuggler's cove from a children's book.

Why invent an island, or any place for that matter? The ultimate blank canvas, where cultures and climate and geography are built from the ground up. A fiefdom with its own laws and rituals; a place to explore politics, conflict and claustrophobia. The word *island* has a power of its own. The visual signifiers it offers – remote, lonely – have the same impact far from the sea. In Mary Lavin's short story, 'In the Middle of the Fields', a widow lives alone with her young children on an isolated farm. Lavin knows the impact of island imagery; its shorthand for disconnection and loneliness and alludes to this in the story's powerful opening lines:

'Like a rock in the sea, she was islanded by fields, the heavy grass washing about the house and the cattle wading in it as in water.'[1]

During the 2012 UK Olympics, a manmade island was towed by tug from Norway to England, stopping at various towns on the southwest coast. Alex Hartley's 'Nowhereisland' – or *Now-Here-Is-Land* – began with the artist's discovery of an actual lost island, which appeared when the Sonklarbreen glacier on the Svalbard peninsula began to melt. Hartley lobbied the Norwegian government to move the whole island to the UK and declare it a nation. They refused, but aware the island would eventually disappear anyway, consented to the removal of a third of the surface material. This formed the basis of Hartley's new island nation-as-installation. Citizenship was open to everyone and over 23,000 individuals applied. The island had its own online constitution, and citizens could propose how it would be ruled – suggestions included banning call centres, free ice-cream every Friday, and that its national currency should be stories, not money. After its UK summer tour, in September 2012 it was broken up. Anyone who had applied for citizenship was gifted a piece of it.

'Nowhereisland' was a situationist utopia that fuelled conversations on climate change, the autonomy of nations and issues of international borders, predating Brexit by four years. It was real but fictional, and only exists now through archive material, and the imaginations of its ardent citizens.

Some decades earlier, US artist Robert Smithson also became obsessed with hypothetical islands. Best known for his land art piece, *Spiral Jetty*, Smithson returned constantly to coasts and water sites in his work. For him, islands were speculative places, and he used various materials and multiple forms – film, drawing and sculpture – to create them. Smithson's islands were circular, represented as either spirals, an island maze or forked peninsulas. One was made entirely of ashes. Mid-lockdown, while trying to finish my island novel, I returned to his 1971 drawing, *15 Islands in a Circular Pond*. While looking at it for what felt like the hundredth time, an email from two artists arrived. Alice Maher and Rachel Fallon invited me to create an artwork in response to a new work of theirs. *The Map* was a five by three metre textile sculpture, of embroidery, appliqué, print, crochet, painting and found materials. It is a feminist map, a political inversion of a *Carte de Tendre*, and part of a series that aimed to recontextualise the life of Mary Magdalene. This map explored the role of women in Irish society and the legacy of religious institutions like the Magdalene Laundries and Mother and Baby Homes. Their invented global space has its own continents, constellations, and several islands, including Oileán Olc (Slag Island), home to Skald's Terrace, Slut Walk and Jezebel Heights. Maher and Fallon's imaginary world collided with my own island and for a while, in the madness of lockdown, of socially distant gallery-meetings, I felt I lived more in these two fictional worlds. For the piece (made with composer Stephen Shannon) I created a soundwork with an Everywoman at its centre; a nomad who roamed this matrilineal universe reporting on what she saw. It ended with a chorus of women reciting the title: *We Are The Map*. My novel, *Hagstone*, features a commune of women who have cut themselves off from society to live on a remote part of my invented island. A year after *The Map*, Alice Maher created her own deck of tarot cards, based on Irish artists in the National Gallery of Ireland. At the opening night, I stood in front of one card with a jolt. On it, were a group of women, dressed identically, standing on a rock in the sea. The card was called *The Island*. Another invented place, but one that spoke to both *The Map* and *Hagstone*'s unnamed setting and characters. A place doesn't have to exist to take hold of us.

For two weeks in 1983, the coast of Miami changed dramatically in appearance. It took years of planning: multiple permit and court applications, commissioned studies on protected wildlife, and an undertaking to clear forty tonnes of rubbish, including car parts, mattresses, beer cans, tyres and fridges. The artists Christo and Jeanne-Claude were undeterred by bureaucracy. After the process began in 1980, finally on 4 May 1983, 430 workers unfurled six and a half million square feet of pink polypropylene fabric around eleven islands in Florida's Biscayne Bay. To add to the artificiality of the scene, the islands themselves were manmade 'spoil islands', the result of channel dredging. The material extended sixty one metres out from each island to cover the water's surface. Free of trash, and encased in shocking pink, they were transformed. They looked orchid-like, almost vulvic.

No piece of writing is an island, in that a piece of writing is not singular. Ideas constantly insinuate themselves into other work. Novels by their scale are multi-tentacled, and it's clear to me now that *We Are the Map* and *Hagstone* – which felt so separate at the time – were tethered to each other, a boat to a jetty, a lighthouse to a rock. Each piece unconsciously informed the other.

When *Hagstone* was in its infancy, I undertook a five day residency on Inis Oírr off Ireland's west coast. Nestled in the Atlantic, it's the smallest of the three Aran Islands, the layout of which Roderic O'Flaherty described in *West Or H-Iar Connaught* (1684) 'as in a sea parenthesis'. The islands were first occupied in 3000 BC, and again in the late Bronze age by people who built Megalithic tombs, which still survive. Its karst landscape of rugged limestone includes drainage systems and caves beneath, a land beneath the land. Getting there took an entire day by taxi, train, bus, ferry, ending with a lift in the diesel fug of a van. It felt like the only place in the world. That it was everywhere *else* that was remote, not here.

I had visited the island several times as a teenager, including a stint at the Gaeltacht, an Irish-speaking summer school. A rite-of-passage first trip away from home that involves eating lots of soda bread and – if you're lucky – kissing other hormone-addled teens. At one end of Inis Oírr lies the Plassey, a 1960 shipwreck made famous in the opening credits of *Father Ted*. A rusted shell coughed up by a wave. Now mottled orange in its final resting place, it resembles the curve of a bracket. At the far end of the island, the lighthouse is an exclamation mark. Lady Gregory, writer and friend of poet W. B. Yeats, visited in 1898, craving solitude and quickly acclimatised, becoming somewhat possessive: 'I felt quite angry when I passed another outsider ... I was jealous of not being alone on the island, among the fishers and seaweed gatherers.'

James Joyce visited the same islands in the summer of 1912, and was captivated by its sense of mysticism and history, from stories of the old kings of Claddagh to sunken Spanish Armada ships lying on the seabed.

He described the biggest island, Inis Mór as 'the holy island that sleeps like a great shark on the grey waters of the Atlantic Ocean'. It wasn't only Irish writers who were drawn to this part of the west coast; Sylvia Plath's only visit here was in 1962 as her marriage to Ted Hughes was falling apart. The couple stayed with poet Richard Murphy, who took them out to Inisbofin, an island further up the Connemara coast. It cast a deep spell and Plath planned to return very soon, writing to her mother in October 1962: 'In Ireland I feel I may find my soul'.

Islands are a place in the mind, a floating piece of grit on the retina, a shape seen in its entirety. An isolated rock is a natural beacon for a writer. Anywhere that's difficult to get to (and to leave), is an ideal place to maroon characters. Nell – the artist in *Hagstone* – uses the terrain as a canvas for her art; the sea and rocks as props. Part of her conflict is whether to stay or go, fearing the end of her practice if she does. Leaving an island is a complicated, time-tabled experience. A reverse pilgrimage for many, who don't look back.

In any story, a writer often asks themselves what is the *one* thing that your characters cannot escape from? It may well be love, bereavement or fate, but if you anchor them to a rock in the sea, it may be the island itself.

AS IN A SEA PARENTHESIS

Last year, I saw a group show at a UK gallery featuring Barbara Hepworth. I've seen many of her monumental sculptures and surgical drawings over the years, and wrote about the latter in *Constellations,* but on display here was a small gold piece that was unfamiliar. Beside it, *Figures in a Landscape,* a 1953 film about her life, played on a loop. Hepworth is shown working by the sea in her beloved St Ives, labouring away with hammer and chisel, coaxing shapes from heavy, raw materials. The camera lingers on the coastline, the sea and beaches, the landscape that heavily influenced her work. And then a realisation, as with the tarot card. That all the years of looking at each brass and stone piece containing holes, I failed to spot the connection between a talismanic hagstone and Hepworth's recurring forms. Shapes whose absence is central to their meaning. In the film's voiceover, Cecil Day Lewis links Hepworth to the land, and her craft to the elements: 'She shapes the stone which water and wind have shaped before her'.[2] A description that also applies to every island, real or imagined.

NOTES

1. Reproduced with permission of Curtis Brown Ltd, London, on behalf of The Estate of Mary Lavin. Copyright © Mary Lavin
2. Quotation from *Figures in a Landscape,* courtesy of The British Film Institute.

Island Fragments

ANTHONY ANAXAGOROU

Vivid memories still find me some thirty years on of my father announcing a summer holiday on the Greek islands of Crete or Corfu. We'd spend two weeks in a modest villa overlooking the sweeping Mediterranean, picking at chilled watermelon, local olives and cheese as we plotted the days ahead. We had no family in Greece, but as my father worked in the travel industry, we were able to visit a few of the islands each holiday. To my mind, places like Crete and Corfu were always islands in the traditional sense – simple, bucolic, gentle-paced and clean, with a seeming abundance of fresh produce. A stark contrast to the country I was born in. A place that although it avoids branding itself as an island still shares its definition. Perhaps the reason Britain prefers not to think of itself in this way is in part due to how exposed and vulnerable islands are often perceived. From a political standpoint, they give the impression of being rudimentary, weak and often lacking in global presence and output. It makes sense that Britain, a country once morbidly obsessed with dominance, sees fit to posture as one fortified by geographic landmass and military might. How one of the smaller of the Northern Hemisphere islands came to dominate 24 per cent of the globe by the early twentieth century will always remain something of a marvel, particularly to those born in and around its far-reaching borders.

As a second-generation working-class Cypriot raised in the suburbs of North London, the intensity of the capital, one aggressively governed by competition and a continuous stream of appointments and pressures, couldn't have been further from those meandering coastlines, idyllic townships and loose, slow days. Despite being only forty miles from the nearest bit of coast, London seemed to me to be a patchwork of municipalities designed to keep people contained and localised. I felt the weight of my own personal history bearing down on me, torn between two competing identities: the British and the Cypriot. It's within this nexus that diasporic people need to somehow balance several notions of selfhood, all underscored by the place a person was raised which is often pitted against the cultural values and traditions of their heritage. The first thing I needed to grasp within the miasma of identity was class and race, and how the two intersect. Language both in English and Cypriot Greek was where I noticed these expressions announce themselves first. My family would code switch between Cypriot Greek and English; the English we spoke was heavily inflected with a strong North London intonation, and the Cypriot Greek was regarded as a lesser, more provincial version of standard modern Greek. Both languages leaned into the lower ranks of society, and both had their origins in empire and colonialism.

Whenever I visit Cyprus, I feel that no matter what direction I take, a body of water will never be far away (unless I find myself in the capital, Nicosia – landlocked, at fifty miles from the sea). When I compare Britain and its varied landscapes to Cyprus I'm often met with a recurring thought: how Cypriot culture is structured around two fundamental aspects: food and water. Anyone who's visited Cyprus will know, the Cypriots take their food seriously; eating is the day's main event and working or washing or exercising are just ways to fill the gaps between meals. Cypriots plan their weekends around beach trips, where they haul ice buckets loaded with an assortment of halloumi and cucumber sandwiches, *lounza*, *koubes* and watermelon. Some go to even greater extremes, bringing along portable barbeques. Such gatherings are made possible by good weather, which cannot help but shape cultures and lifestyles, informing an island's relationship to water. The same can't really be said for Southend or Edinburgh, or Brighton or Bournemouth, the British weather curbing year-round beach activities. I've spent some amazing summer weekends around the Seven Sisters cliffs in East Sussex and on Canvey Island in Essex, but they have never matched the Cypriot experience, one surrounded by sea, under a blazing Levantine sun.

When both sets of grandparents arrived in London from Cyprus, they were, on paper, already British citizens. Both my parents were born in Cyprus under British rule which automatically 'granted' them British citizenship. Indeed, when they recently applied for a Cypriot passport, the authorities concluded that they were in fact not Cypriot but British, as at the time of their births in the 1950s, Cyprus was still part of the colony.

My father's mother would arrive in London and embrace Britishness by teaching herself a high standard of English both written and spoken. She would involve herself in current affairs and global politics, adopting popular values in the hope it would help better her chances of moving up the social ladder. She dyed her hair blonde after it turned grey, introduced herself as Maria instead of Maroulla, and could regularly be found debating the various policies of Britain's political parties. As a lifelong Tory voter, her last ballot paper in 2022, before she died, went to Labour. My maternal grandmother went in the opposite direction. Choosing to inoculate herself from British society, she worked only for other Cypriots in North London's rag trade. She would buy her daily produce from Cypriot greengrocers and over time carved out a social network that consisted entirely of other Cypriot women. Often when we'd talk about her life, my grandmother would tell me that she was a Cypriot: not a Greek nor a Turk, nor a Brit, and furthermore she was a socialist not a nationalist. Both women identified as Christian Orthodox yet understood God in different ways. My mother's mother was semi-literate, yet my father's mother happened to be one of the first women in Cyprus permitted by the British to attend secondary school in the 1930s.

I was twelve years old when I accompanied my father's mother to her local butcher's in Dalston, Hackney. We stood in line waiting for a man in a bloody apron to take our order. 'Yes sweetheart,' he called out, 'what can I get you?' She pointed through the fridge glass at the cutlets of raw meats, specifying weight and her preferred joint. The butcher, somewhat flirtatious, causally asked where her accent was from. 'Greece,' she replied. I called out instantly to correct her, saying, 'No we're not, we're from Cyprus.' She squeezed my wrist and from the side of her mouth told me to shut up. 'Greece,' she said again, 'We're Greeks ... but from Cyprus.' I spent years turning that one incident over in my head, asking myself why she chose to identify with a totally different country over the one she was from. Why was she ashamed of her Cypriot heritage and how did she think others saw Cypriots?

One evening in 2019, at a reading in central London, a member of the audience accosted me to ask where my surname was from. I told her I was Cypriot. 'Greek or Turkish?' she snapped back. I remember thinking that my paternal grandmother would have said Greek, then gone on to explain how the Cypriot people are actually closely related to Cretans and how we are all descendants of ancient Greeks, closing with a quote from Aristotle. 'Just a standard issue Cypriot,' I said again. She looked confused, as if my ethnicity didn't exist, or I was being deliberately evasive or facetious. I thought about how, historically, nobody, including the British, has ever really known who Cypriots were, nor how to tell the different ethnicities apart. When the British took the first Cypriot census in 1881, they recorded that Greek Cypriots made up 74 per cent of the overall population, with Turkish Cypriots constituting

24 per cent and the remaining 2 per cent being Afro or Armenian Cypriot. As the two core groups looked so similar it proved incredibly difficult for the British to distinguish between them. Sometimes it was a case of noting the type of headscarves the women wore, other times the subtle difference in how particular words were inflected, but there was never enough certainty even in these distinctions. People largely spoke Cypriot Greek, and the Turkish Cypriot communities knew enough to converse with their neighbours and vice versa. Eventually, religion became the only way the British could categorise people. This led to many Armenian and Maronite Cypriots (who were Catholic, not Orthodox) being considered Greek Cypriot, and Turkish Cypriots, who were mostly secular, being labelled Muslim.

This categorisation, somehow at once both limiting and slippery, has plagued the Cypriot identity for most of its existence. This is precisely why the nationalists in the 1950s and 1970s were able to weaponise and harness Hellenism and Turkism, in the hope of finding a more tangible Cypriot nationality, one that was wholly their own. This specific imprimatur signed off by right-wing groups in Greece and Turkey has travelled down the generations of my family and gone on to haunt me for most of my writing life. It's a clear hangover from the colonial era, underpinned by an island's existential fear and desire for independence which finds expression in a preoccupation with ethnonationalism and Pan-Hellenic hegemony. This process of diminution, of erasing the Cypriot and privileging the Greek or Turkish signifier, is a form of both ethnic and cultural erasure, with the dominant and classical narratives of Greece and Turkey supplanting those of Cyprus along with its multivalent and distinct history.

My father understandably inherited similar strong Hellenistic ideas from his mother. His argument suggested that to be Cypriot was muddy and deemed less sophisticated than to be Greek and that Greek Cypriots do in fact share a religious, linguistic and cultural affinity with Greeks, not to mention how 1,400 years ago Cypriots would have come from the island of Crete via the Minoan civilisation, followed by the first wave of Mycenaean Greek settlers after the volcanic explosion in Santorini. Throughout my teens and into adulthood I'd rebuke his claims, defining the Cypriot as belonging to their own racial and historical category, asking who the indigenous people might have been, arguing that the Mycenaeans were ultimately undertaking the work of colonialism if we agree that there was already civilisation on the island prior. As these arguments between family members dragged on through the years, I became more determined to regard the Cypriot as someone distinct from the antiquated pageantry of Greece or Turkey. We were as Greek as a Brazilian is Portuguese.

The first time in Cyprus's 2,500-year recorded history that it was allowed to be independent from foreign rule or interference was 1960. And even now, Britain possesses 3 per cent of the island, for use as army bases from which to launch military campaigns across the Middle East and North Africa. Over its millennia of history, Cyprus has been governed by nearly every major power located within Asia Minor, sometimes over entire centuries. On the one hand, the island's geographic location, situated in the heart of the Levant where east borders west, makes it highly sought after; on the other hand, its topography and small size deems it constantly exposed, and susceptible to attack. An anxiety still felt by many Greek and Turkish-speaking Cypriots, who remember first life under British rule and then the subsequent turmoil that resulted in civil ethno-communal unrest, followed by a Turkish invasion and illegal occupation (as determined by international law) of which just under 40 per cent of the island was assigned to Turkey, and ratified by both the British and Americans.

Given how identities are constructed and how people come to view themselves through a geopolitical lens, it makes sense that my late grandmother would want to see herself as Greek rather than Cypriot. Or why so many Cypriots in London will refer to themselves as Greek or Turkish before Cypriot, leading with nationality rather than ethnicity. Part of the issue here is that when Cyprus was looking to break away from British rule in the early 1950s, nationalist groups like Ethniki Organosis Kyprion Agoniston (EOKA) wanted to rid the population of the idea of Cypriotness, which to their mind was too vague, too diffuse with other ethnic denominations. Instead, they, along with the Orthodox church, drove Hellenism into the minds of the Greek-speaking populace. It was during this period that both my grandparents would have been in their late teens, where their politics and worldviews started to take shape.

My professional bio states I'm a British Cypriot writer. I often ask myself: why does this matter? Why can't I just be a writer from London? Do I overidentify with my ethnicity? My bio is read in an anglophone vacuum, where I'm directly in conversation with ideas of what it means to be British and Cypriot. To straddle two islands, two worlds, two experiences at once, is to draw upon a duality that informs the way an individual arrives to organise the world. I'm still wrestling with my late grandmother's desire to see herself as something she wasn't – Greek and British, white, Christian and Conservative – with the internalised shame many elder members of my family felt when arriving in Britain, of having to tell people they were from an impoverished colonial island, ravaged by war and oppression.

One holiday, on the Greek island of Spetses, my father relayed how when he was young he would tell people he was Italian or Spanish, as Greeks (by which he meant Cypriots) were very much an undesired group in the London of the 1970s. One evening, during their early months of dating, my parents, along with my father's brother, were walking home through Kentish Town when two tinted-out Mercedes pulled up beside them and half a dozen white men stepped out, armed with crowbars and glass bottles. They attempted to sexually assault my eighteen-year-old mother and viciously attacked my father and his brother, hospitalising them both. When we discuss the attack now, my father is still convinced it was motivated more by xenophobia than racism, because to accept he wasn't perceived as white, or English or British, would mean having to reorientate his whole understanding of how he sees himself in Britain. My mother, who is far darker skinned than my father and grew up with a mother whose idea of being

Cypriot was almost antithetical to that of my father's mother, still considers it a racially motivated attack.

I often contemplate this event and how it perhaps became the fulcrum for how I navigate my two islands. I've learnt to straddle their shared and complex histories – the presence of the British in Cyprus is an equal part of both countries' histories. The seacoasts of England and Cyprus with their sometimes-violent nature confounds the linear maps of time. Their people respectively look for clarity among the cultural fragments and tensions, a clarity that perhaps never existed, yearning to understand the past. Cypriots, much like the British, have since become obsessed with a kind of racial purity; perhaps the only thing they now have in common.

My father is now in his seventies. Last summer we sat in his garden talking about how much Cyprus has changed over the course of his lifetime. He laments how the island has tragically compromised so much of its unique character so as to pander to foreign investment, presenting as another strong European country, more western than eastern, denying its geography and heterogeneity in the hope it will appear less vulnerable; less island, more solid landmass.

I bring up the idea of retirement and living out the rest of his life back in the place where it all started.

I tell him London is dead now; it's no longer the city it once was. He stays quiet for a while, taking another swig of beer, checking the football scores on his phone, before saying that Cyprus is where it may have started, but England is where it will end. My mother and her sister come out to join us. We talk of an autumn trip to Cyprus for a week or two. There the island will remain; cushioned by water, pinned down by heat, no matter what's happened or what's yet to come.

A Portion of Land and a Cow

ORLAINE MCDONALD

How often do we consider the island-ness of the UK? I rarely do – or rarely did. Political butchery in recent times has forced many of us to reckon with this thought in some form or other. But we are, of course, very much a distinct mass of land surrounded by water.

I was born here, smack bang in the middle of it, the Midlands in fact, miles from any lapping shoreline. I hold a UK passport and England is the country of my birth. Yet I am born *of* Ireland and Jamaica – a distinction as weighty as the combined land mass they signify. I am born of Ireland and Jamaica through the bloodlines of my mother and father. They in turn born of immigrant parents who chose this island to settle.

Ireland, Jamaica, England: I am none of them and all of them.

A few years back, I found myself at a pub lock-in. A beautifully un-gentrified pub in Southeast London. I was the only woman and the only person of colour present among a select group of mainly Irish labourers. My gender, and the colour of my skin, did not initially draw any attention, although my new friend, with kind eyes and weathered hands, leaned in and reassured me I was one of the boys now, and in luck because soon the big man himself would be gracing everyone with his presence. Big man? I asked. Him upstairs, he confirmed with a grave nod skywards. God? I squeaked. It was fair to say we'd all had a drink. Big John, he whispered, reverently, the landlord.

Big John emerged from the sway of stripped curtains between the cash till and optics with the regal bearing of both Pope and pop star; tall, wide and wearing an impressive sheepskin coat. A space was cleared at the bar and Big John perched his spectacular frame on a barstool. The atmosphere hushed as pints were studiously lowered. Punters shyly, hesitantly, approached to shake his hand. Some were pulled swiftly into a mighty, violently hearty embrace and others, inexplicably, were dismissed with a simple gruff nod. Talk, both political and ponderous, resumed, lower now; all energy in the pub sucked and syphoned into the air around Big John like a pool of golden light. When his eyes came to rest on me, I was beckoned with his meaty slab of a hand. My new friend gently propelled me forwards across the patchy carpet.

Big John considered me with narrowed eyes. Asked my name, commented on its unusualness, closing his eyes, nodding slowly, as if meditating on it.

'And where, are you fram, *Orlaine*?'

'Sydenham.' I whispered.

'Where are you *fram* fram?' He leaned in, 'Where are your *people* fram?' Eyes still closed, head tilted towards me, deeply frowning.

Ah. The question.

'My father is Jamaican, my mother is — was, Irish.'

My new friend jumped in, unable to contain himself.

'Castlebar! Her people are from Castlebar!'

Big John's eyes flew open. He brought his fist down hard on the bar. 'Get this woman a *drink!*' He bellowed.

It was a strange celebration, this acceptance, this acknowledgement of one strand of my bloodline.

Growing up, England was *a* home, but not *our* home. Home was *back* home: Ireland, Jamaica. My parents conjured up these places potently. Bog, outcrop, Comorant and Guilliomot merging seductively with mangrove mist and heat, Mourning Dove and Humming Bird. Mythological wonderlands baked hard into my soul.

My mother was the youngest of seven, five brothers and one sister. We cousins score many, all of us, throughout our childhoods regaled with family lore. When we compare stories, some are similar, some conflicting, but always there is cold, often hunger. Shoes, the lack of, a recurring refrain.

My mother would often say, *it's the Irish in me* to explain some idiosyncratic tic. She draped her Irishness around her, a comforting cloth of self. This Irishness, I came to understand when I was very small, was the reason she did not look like me. Why her eyes were the colour of a stormy sea, why freckles danced across her shoulders when she danced, why the sun coloured her red not tan. She fed me wild stories of my grandfather, Jack; a rogue; a drinker; a wife-beater; a working-class intellect. In her teens my mother swam at competition level only because he threw her in the deep end of a swimming pool when she was eighteen months old and instead of drowning she lived. He claimed to have joined the IRA because all the young men in his village did. He also joined the British army just to get a pair of shoes. He held no fealty to either, according to

A PORTION OF LAND AND A COW

my mother. He was a tailor who made suits for Saville Row and would routinely write *cunt* in the lining of those he made for royalty.

My father's memories of Jamaica were snapshots, a flickering reel of shuttered light. Here he is peering through a glass-bottomed boat to watch skipjack glide underneath. His father striding down to the shoreline, my father atop his shoulders; his father's towelling swimming-trunks, fire-engine red, his father swimming deep, my father, naked, paddling in the warm shallows. See the scar on his father's face, a permanent souvenir from a job opening barrels of acid, and the curtain dividing living from sleeping, see the baby there nestled in a drawer. And of course all the leavings, the quiet underscore to it all. Left for good when he was eight, the baby four, so my grandparents could journey here, the Mother Land. And when my Grandmother discovered the people she'd entrusted her children to, sending *money* to, were falling dangerously short, she sent letters with instructions for a better foster family to be found, hoping and praying, parenting as best she could from across the sea.

My grandparents on both sides sold something treasured to pay for passage to this island. In Jamaica a portion of land, in Ireland a cow.

There is pain and loss embroidered through my parents' remembrances, passed down from their parents, and when the pain and loss is unspeakable, there is silence. There is recklessness, and deep pride and rebelliousness, because England was not a welcoming place for either.

No blacks, no Irish, no dogs

Before I knew anything, I knew that we were here because the lands of our origin were scored by the blood and violence wrought by the English who feasted on our oppression. This is what my parents taught me.

Yet my relationship to these places is complicated. Despite a childhood steeped in the heady romanticism of Ireland and Jamaica, we never went back home. We never returned to those fabled lands. We never even visited. I had not touched the soil of either. Those islands, woven into my very fabric, were unreachable. I no more belonged to them, than they belonged to me.

Now I understand, of course. My parents were too busy getting on with the business of getting on, getting by. Living was day to day. Money for travel was a flight of fancy in a world where my mother used loan companies to pay for Christmas and sometimes just to keep the lights on.

But still, back then, this failure to truly know *back home* scratched at me, marked me with indelible shame. This added to the other shame, a steady stream of it eddying down the neck of my young self. Of being mixed-heritage, brown-skinned. A child raised by a white mother in a white town. Painfully hyper-visible yet also invisible. Too much and not enough.

And whether I liked it or not, as a child in the 1970s suburb of Birmingham where we lived my brown skin marked me out as someone who did not belong. When I was five there were no other brown-skinned children in my class. I stood out for being undersized, a tiny brown thing with too much hair. During story time our teacher would sweep me onto her lap and cuddle me as she read, stroking my skin and hair. The other children, cross-legged on the floor would stare up at me, confused and resentful.

Later, in another school, in an even whiter town, a teacher made me stand up and explain to the class what yam was, and how *you and your family use it to cook*. Only I couldn't because I had no idea what yam was. Her irritation was palpable.

Then an inexplicable occurrence: at six years old, I go with my mother to visit my Irish grandfather, the wild anti-hero of a thousand myths, now an old man living in a prefab. Him pointing at me with a tea towel in his hand, ordering my mother to *get that pickanniny out of the house.* Waiting patiently and obediently in his overgrown garden until the visit was over.

Shame heaped upon shame, a creeping stain, a heating up, an urgent desire to flee or pee.

If here is a place that doesn't want you, and there is a place that doesn't know you, then where is home?

Once, newly divorced, in the early days of a cold April, I decided to book a break somewhere warm and cheap. I chose Gran Canaria. A small resort on the southeast tip of the island. The keys to the apartment were handed to me by a man who glanced from me to my small suitcase and back again, before murmuring something in Spanish. The place was tired, basic, but clean. A living room-cum-kitchen area in one room, a bedroom in the next with two twin beds, and a small, blue-tiled bathroom next to the front door. The balcony was huge, spanning the entire width of the apartment. A white plastic sun lounger, table and chairs had been placed in the middle. A wall divided each balcony from its neighbour. To my right, a pair of male navy swimming-shorts were draped on a chair, along with a small greying towel – the owner never to be seen. I had the distinct feeling I was the only person there and hovering on the horizon, glinting like a plate of jewels, was the sea.

A PORTION OF LAND AND A COW

I remember making my way through the desolate complex. I stood on the dusty, carless road and let the relentless sun hammer ferociously at my bare head. I followed a weaving path down through a garden flanked with tall pine trees.

The resort was nestled inside a curving cove. Buildings tucked untidily upon the looming mountains; there was no order, no discernible pattern, just white geometric structures, with rows of balconies and windows jutting out. Pigeons, it seemed, were the only type of bird.

I felt the pull of the sea but also an urgent need to return to the apartment. That familiar feeling of exposure. My Blackness, my aloneness, marking me out as some kind of oddity.

I sat at a beachside café, used my limited Spanish to order a large rosé, contemplating the teal colour I'd painted my toenails; the exact shade I'd chosen for the new tiles in my new kitchen, in my new home. A bold choice, people commented at the time. As if by extension I too had become bold. I was not. I was simply learning how to be husbandless, and because past losses tend to attach themselves to fresh ones – collecting and swelling to outlandish proportions – I was also learning, again, what it was to be motherless. Choosing tiles, in comparison was an act of comforting banality.

On the first night I met a man in a bar. An Irish man. We drank and we talked and he spoke of his life, his wife and his children, with such candidness and tenderness I was intoxicated, and after we said good night, as I watched him slope off into the thick night, arm raised in salute, the feeling carried me back to my apartment. I was drunk on him, the warmth and the smell of pine and sea sifting through the dusty gravel. My movements were thick and languid, my thoughts expansive and ranging. In that moment I was less alone. It felt revelatory, heady and delicious. The possibilities for anything seemed endless.

On the second night I met another man, a Spanish man, and something happened. No. That is not true. On the second night, the man who owned the keys to the apartment appeared at the door and I let him in, which is worse, but true. And something happened. But the thing could not be recalled, not in its fullest sense, and so I was unsure how much I had invited it, because hadn't I invited him in? The not remembering was a problem, it made me feel complicit. It left a void, a blank space, and in that fertile space, shame crept in, familiar and eager. So, to escape the thing, the shame, I packed up my small suitcase, left the apartment and paid a taxi to carry me north, to Las Palmas on the opposite side of the island. I booked myself into a hotel of glass and marble, squarely facing the sea. One night, watching waves leap out of the indigo, two thoughts occurred; firstly, I appeared to have arrived at middle age pathetically unmoored, followed by the far more agreeable: it is always good to have the means to escape.

I have tried to process the events of this trip into a short story. It doesn't work. I tell myself it's because it contains too much of me and not enough. But the truth is, shame remains.

A wise woman once reassured me that nothing stays the same. I love this woman deeply. In the absence of my own she has mothered me well.

She is right. Less and less I flounder in the seas of my own recklessness, and less and less I swim towards Ireland and Jamaica for validation, sense of self, belonging. I am more than my dual heritage; proud, yes, but more than any one island. Age? Perhaps. Or maybe just the business of living. Perhaps while navigating life's slippery contours I've somehow managed to come to a deeper understanding and acceptance of my place in the world, where I fit.

I am a grandmother now. When my granddaughter lies beside me and faintly, briefly surfaces from her deep sea of sleep, her small limbs rearranging themselves, and she reaches for me, arms flung heavily around my neck, I know that in these moments I am exactly what she needs. My body the landmass she is seeking. Yes, in these tender, bruised hours before dawn, floating unapologetically regal and serene, I am home.

Island Hopping

NICOLA DINAN

I was born on an island – Hong Kong Island, to be exact – but my parents moved me and my two sisters to Kuala Lumpur, Malaysia when I was six.

I was the youngest and least popular of the three. I spent my weekends playing World of Warcraft and having long silent Skype calls with my best friend while we finished our homework. Until I was fifteen, my family lived in a residential compound on a golf course, forty-five minutes outside of the city, and far away from opportunities for misbehaviour. Monitor lizards sunbathed on our driveway, and monkeys who had claimed the empty plot of land across the street stole bananas from our kitchen. My sisters snuck out of the house at night, taking taxis to the bar strips in

Bangsar or Changkat. They came home with hair reeking of cigarette smoke, armed with bags of McDonald's – my bribe for keeping their escapades a secret from our mother.

I was, in other words, a neek. I was anxious, hated getting into trouble, and prized doing well at school above all else. Once, after a particularly difficult but routine physics test, I cried for days, uttering the words 'I would be happy to die' (I got an A). I now understand that the storm brewing inside of me was simply a part of growing up, though perhaps in my own special way. Becoming a teenager had brought an immense complexity to the ways I'd begun to understand my body, sexuality and mind, all of which seemed to exist purely to antagonise me. Control became my modus operandi, and the OCD that had started to germinate in childhood bloomed ferociously. I controlled 'bad parts' with *good* behaviour, and I tempered my desires with shame.

And so it was to the surprise of everyone around me – even myself – when I was suspended from school at thirteen during a trip to Pangkor Laut.

I remember feeling so excited about the beer, which the man running the hotel shop had happily sold to my friends and me. We sat on the thick grass behind our hotel – consisting of small beige bungalows surrounded by a vast expanse of trees – bringing the sweaty bottles to our lips. When our teachers inevitably found us, they tracked us down in our rooms after our poor attempt at running away, and sent us home for the remainder of the week.

Then came a meeting with the school principal, who refused to let me sit as he berated me for my irresponsible behaviour. I went home doused in self-pity, ready to light a match. My mum, who flew home early from a work trip, couldn't even be mad: I was so clearly ashamed of myself, and would spend the next year ruminating over my bad decisions. *What* came over me? What *possessed* me to behave this way?

And yet, this did not stop me. I would continue drinking on school trips for years to come, even at a Model United Nations trip for geopolitically minded teens on the island city-state of Singapore, where we tried our luck with flimsy fake IDs. On my final school trip – a diving excursion to Pulau Perhentian – two friends and I drank with our much older diving instructors, who had questionable intentions.

'We don't know who these men are!' my teacher screamed at us. 'They could be dangerous!'

She was, of course, correct.

There was something about islands. A certain, largely false, island logic – a vision of life without its regular constraints. While kids my age in the UK went on trips to Ayia Napa and Zante to celebrate the end of exams, my friends and I went to Phuket and Koh Samui. On one of these trips, our group were ushered into a ping-pong show, only to watch a woman squat over a cup and release a terrapin from her vagina. This, for many of us, represented *some* kind of line, but we walked out of there without without sparing much thought on what might have led a woman to endure such pain. At the Half-Moon Party on Koh Phangan, some of us tried mushrooms with little awareness of the consequences of drug use in a country like Thailand. After all, we were foreigners and, more importantly, some of us Westerners – people who, even if subconsciously, believed the rules did not apply.

Maybe it's that island logic equates islands with freedom. After all, it was on one of those island holidays where I first came out to my friends. My family had known for a while, but I waited for island logic to take hold before confessing the same to my peers. On an island, restraints loosen, harder fabrics melting into soft and stretchy polyester and nylon.

Before I was a writer – or before I allowed myself to *call* myself a writer – I was a lawyer at a Magic Circle firm. During my time there I felt as I have often felt: misplaced. Working there defined the better half of my twenties. I thought it might clean the muddy lens through which I'd always viewed myself, work a *good job* and life will fall into place. However, the promising opportunities that it offered – the large salary, most of all – only seemed to soil the glass further.

Many slews of late nights and competitive asslicking earned me a six-month secondment at the firm's Hong Kong office. It felt like an odd sort of homecoming. I was and still am a permanent resident of Hong Kong, a benefit of having been born there, living there and having Chinese heritage. Still, I was, back then, undoubtedly a foreigner.

Though I'd visited after my parents moved to Malaysia, I'd left Hong Kong when I was six years old. I would be returning to the island as an adult without a lick of Cantonese. My flights there and back were already booked, my rent for my twenty-third-floor apartment with harbour views was paid for, and at the start of every month, one of the office secretaries brought me an envelope with a wad of Hong Kong dollars enclosed – my monthly pocket money.

I was a twenty-four-year-old on a six-month island holiday. I worked in a tower by the harbour, a metal box punctured by nautical windows that looked out towards the sea. As my two friends and I were inducted in a meeting room, I stared at the sky and ocean, two unblemished Rothko panels. On the one hand, I was on the precipice of a new beginning, and on the other, I was an expat.

Expat is not a compliment. Expat behaviour is certainly not a mark of grace and elegance. This remained true for nearly all the time I would be in Hong Kong. I blacked out at a festival by the harbour during the set of author-cum-DJ, David Irvine. I spent Sundays at bottomless sushi brunches, watching my peers stand on tables and dance with impunity, shaking my head while double-fisting glasses of bubbly. I obliterated my senses on the company junk boat, throwing up from a combination of motion sickness, carb-bloat and alcohol consumption. As I stumbled back to my apartment, I would often stop by the 24/7 McDonald's right next to the lobby. My order was always two, not one, Filet-O-Fish burgers – à la carte, no fries, no drink – to eat in bed, throwing them up right after.

This island – one of Asia's financial centres, but an island nonetheless – was yet another site of misbehaviour. My friends and I were tolerated for the money we spent and the institutions to which we were tied. We were toddlers when the handover of Hong Kong from Britain to China took place. And yet it seemed that to many of its visitors, including us, this island would always be a colony, a playground on which to stomp.

By the end of my time in Hong Kong I had started to become reclusive, declining dinners with secondees from other firms. I stayed in my apartment eating surprisingly affordable takeaway sashimi while playing games on my phone. I began to count down the days until my return home – and until then I'd scarcely considered London home at all.

It had become clear to me that island life, or at least the suspension of reality in favour of hedonism and one's own rules, could not sustain itself, at least in my body and mind. During this period, I tumbled into old habits. I took measuring tape to my waist, charting the loss of inches as I sacrificed carbohydrates, a deliberate punishment I'd enforced on myself as a disordered penance for my heady indulgences. I regressed into my longstanding hypochondria, abusing my health insurance for extensive tests on my body for issues my neuroses spawned. I took taxis up to The Peak – the very hospital where I was born – for examinations, X-rays and stool testing.

My weekends became quiet. I would take the thirty-minute ferry to Lamma Island, walking through the tight corridors of Yung Shue Wan, glimmers of the sea poking between the narrow pockets of air between buildings, seafood restaurants hanging over the shore. I'd pick up a cone of fried chicken, eating half of it as I watched the waves roll against the big power station, and the other half while walking the trail to Sok Kwu Wan on the other side of Lamma. I'd take the ferry back before sunset, planning for what my life would look like when I returned home – the big changes I would make, the care I would start to give myself.

I left Hong Kong in 2019, just before protests erupted over the proposed national security law, which would criminalise the behaviour of dissidents and open channels to extradite suspects to China for trial. University students formed barricades. Some, in black hoodies, shot flaming arrows from rooftops, while others traded petrol bombs with police, who in turn were armed with rubber bullets and tear gas canisters. In Sai Ying Pun, the area where I'd lived, protestors defaced the Chinese national emblem at the China Liaison Office.

I was nowhere to be seen. I watched from afar in London, my mother forwarding me videos of the protests, the damage and scathing commentary from her friends. Some decried the use of violence, most mourned for the Hong Kong they'd known. Many, unsurprisingly, were expats who'd enjoyed the spoils of those final stretches of colonial rule, and had long left the island.

The protests delivered the moment at which island politics would irrevocably cleave through island logic; fun and games a thin cover for what rumbles beneath. I watched videos of smoke billowing up towards the apartment block I'd lived in, but I would not be there to smell it. In the face of Hong Kong's shaky future, many companies and expats engaged in a practice I've known since I was a teenager: island hopping.

Eyes turned towards Singapore. Hot and humid like Hong Kong, but all year round. The land flatter, the buildings just a little further from one another, their foundations stretching wide into the soil. Financiers with European or North American passports brought their families and lives in tow. They took the cable-cars in Sentosa, watching a skyline of swelling rents. One city deflates, the other bloats. Another long vacation, convenience always. *Hong Kong is dead! Hong Kong is gone.* But to hop does not mean to crush – the land persists, the earthy expanse taking large, deep breaths.

Love Island; Dating Manhattan

MEGAN NOLAN

My friend Jay looks a little like a caricature of a New York eccentric. He has the kind of glasses and straggly hair which cause people on the street to occasionally exclaim 'Oh, John Lennon!', and is the only person I know who still smokes cigarettes continuously through each day and night. He is rail thin in a way which could suggest either aristocracy or scurvy; when we go to dinner he doesn't mind much what he eats as long as he can have an extra-cold dirty vodka martini and an espresso simultaneously to begin the meal. He dresses beautifully in snug, well-worn wool jackets and Victorian looking shirts and trousers that fall like somebody made them just for him, which they sometimes have. Jay lives in Brooklyn but is a Manhattan guy at heart.

He works sometimes installing paintings in the homes of enormously wealthy city people, and when he's done for the day we go to dinner at The Odeon in Tribeca then walk to the East Village to drink and talk. Jay doesn't live in Manhattan for the reason hardly any of us do except a particular genre of beautiful young women who seem to stumble into rent-controlled Chinatown setups with nothing but a suitcase full of hair ribbons and a ragdoll cat on a leash. Or the guys in their thirties and forties who are content to share austerely outfitted apartments with roommates, for the thrill of being able to say they live in Midtown; apartments so devoid of characteristic and aesthetic flourish that they beg the question: What if Patrick Bateman was broke?

I took a photograph of Jay on the morning of 17 March 2020, standing naked in his elaborately packed studio. Back then he wasn't my friend yet, he was just the last guy I might ever sleep with. This was not because we had fallen in love, but because the world was ending. I had arrived in New York a few weeks previously, planning to stay for three months but maybe forever if I could swing it.

Before 2020 I had been here only once before, on a four-day press trip where I interviewed an exhausted teenage TV star in the basement of an opulent Soho hotel. That trip was given to me as a kind of consolation prize by the London magazine I worked for after they declined to hire me on a permanent basis, the upshot being not only a free flight to New York but also staying in said opulent hotel with a $250-a-day room service allowance paid for by the film studio promoting her first feature. Even a nightly steak and red wine took up only half of that, enabling me to ostentatiously give a 50 per cent tip to the porter. When I had my slot with the teenage TV star, who was known for her outspoken views on matters of social justice, she seemed almost concerningly void of the will and ability to converse, which was no surprise after having to speak to journalists in twenty-minute increments continuously from 9 a.m. to 4p.m., which it now was. No matter what I asked, she seemed to circle back, with a thin watery smile, to the fact that she liked grilling on the beach with her mum, who, along with her manager, sat a few inches away from us. The piece would be terrible, I knew, bland and rote. Afterwards I stepped out into the sunshine, bought a hotdog and sat by the water and cried with happiness anyway.

My original three months in New York were a trial run to see if I really wanted to move here; although I already knew I did. It just seemed difficult to admit this on the basis of one visit lasting less than a week and a lifetime of fantasy constructions half-recalled from television and popular culture. I had also come because, for the first time in my life, I had more money than I needed, having been paid a to me unthinkable amount of it for my first novel. I would later confess the exact figure to a finance man I had been forced into conversation with at a party in Soho by a playwright with evil in his heart, who told me the finance man was a kind of benefactor of his; I would tell the man that though I had dreamed all my life of having enough money for restaurants and expensive dresses, the minute I got it, I was so frightened and uncomfortable that I immediately wanted to drive it away from me – which I did, with alarming alacrity – either by insisting on paying for large group dinners of acquaintances I hardly knew or liked, or by donating chunks of it to charities on a whim. When the man asked me what the figure was and I told him, he laughed mirthlessly and said, 'Honey, I make that in a week,' and so I told him what my annual income was before my book sold and he barked with horror and said, 'Honey, I'm making that right now, while we *speak*, while I'm not doing *shit*.'

I had also come to New York because I wanted to be single here. All the ways people described New York as being a terrible place to date sounded just fine to me: the chaos and the transience and how nobody means anything they say. I was going, eventually, to live in Windsor Terrace, a Brooklyn neighbourhood traditionally populated by Polish and Irish immigrants and now one of the increasingly bougie park-adjacent media-worker-parent enclaves. But before my real life here could begin, I checked into a hotel in the city for a week – as close to Central Park as I could afford – and began to meet men. I went out with a comedian called Eli and brought him back to the hotel, where he took a break from going down on me to look up and say 'I hope this is what you thought Manhattan would be like ... getting eaten out by some bald guy'. In the morning, we ordered grapefruit and coffee and he asked for cream with his.

'It's so indulgent,' I marvelled, 'So excessive, to have actual full cream in your hot drinks.'

'Welcome to America, baby,' he grinned, and winked. Eli became my friend. Back then, he lived on the Upper West Side but now lives in Brooklyn with his girlfriend. I saw him last night, in Chelsea; we sweated on the sidewalk holding our tiny fifteen-dollar vodka sodas outside a reading, the erotic breeze of the first of May surrounding us while I told him about the person I was in love with.

Then there was John, a writer. John and I met on Tinder and texted while I sat in the window of a bar feeling a little sad. 'What are you doing?' he asked, and I replied that I was doing nothing because a date I had been on was bad and I had left early. 'Well, do you want to go on a good date instead?' he asked, and then showed up at the bar thirty minutes later. John, too, remains my friend. He lived in Brooklyn then but has now moved to the Upper East Side – unlike the rest of us – John is ascending. He is invited to embassies and appeared on the former President's summer reading list (jealous rivals have since drunkenly disclosed how annoyed they are by his success – they lean in and ask rhetorically 'Do you even think Obama *reads* the books he posts about?'). As a New York native, John's entitled to the city, and it suits him. Like a boorish character in a 1990s sitcom, he now behaves as though Brooklyn is a distant and unthinkable location and once described himself as a 'Manhattan chauvinist'.

Before I left London, Covid was still a thing people were joking about. Having stayed up all night at my leaving party singing karaoke and smoking cigarettes, I developed a hacking cough, which my friends teased me about taking to New York: super-spreader, patient zero. I refused to internalise the pandemic as anything relevant to my life until the day they announced the bars were closing. They had already advised against

LOVE ISLAND; DATING MANHATTAN

kissing, and touching your own face. Now the possibility of dates was fully extinct. When it became clear that the pandemic was not going to wind down conveniently, I booked a flight out of New York for the following day and packed and left to drink wine in Jay's studio. We had agitated, strange sex which I would think about for months afterwards, being as it was the final intimacy I would experience for a long time.

The day I left I first walked from Windsor Terrace all the way uptown, fearing I may never again see any of it. I crossed the bridge, where some tourists were still half-heartedly taking photographs, and the sky was fittingly ominous, draining the Manhattan skyline of its usual jaunty splendour. Instead, it had become Gotham proper, looming and menacing. When I walked over into Chinatown I sat and smoked a cigarette on a bench outside a juvenile detention facility and a cop in a mask said 'You can't smoke here,' and I shrugged and he shrugged back at me and walked away. Later that night I took a taxi to JFK. My suitcase exploded as I tried to get it in the car; I taped it up while kneeling on the airport floor, weeping silently. I had been in the city less than one month.

Once life seemed largely possible again, I began to return, visiting for longer and longer each time, until

I finally admitted to myself I would have to try really living in New York. I needed to be convinced that it was impossible for me to survive here, in order to really accept anything else. I schlepped all over the sublets of Brooklyn, and for one boiling August I lived in Manhattan, going crazy in a cupboard sized studio in Gramercy Park. During this heatwave I did a reading in a packed gallery in the Lower East Side where people spilled out onto the street crouching down on the filthy sidewalk to press their faces against boxes of cold beer. There I met a man named Grey; he and I would date for a little over a year. Manhattan was important to our conception of ourselves as a couple, particularly as I was figuring out who I would be in my new life. No wonder the city was wrapped up in all that mythology and bullshit; New York is the most wonderful place in the world and it is also built on a decaying foundation of suffering and filth. The glamour is plastered over revolting inequality and the persistent lie of meritocracy, of the American Dream. Likewise, I really did love Grey, we did share some of the most unforgettable experiences of our lives flitting around Manhattan, drinking twenty-dollar cocktails and admiring ourselves in the restaurant mirrors, but they weren't rooted in something lasting or real, and when the glamour wore off there was nothing left to hold onto.

LOVE ISLAND; DATING MANHATTAN

On our third date we met at Grand Central Station and went to the Oyster Bar, and later he would tell me that when he saw me approaching from a distance wearing Annie Hall-ish loose cream slacks and a men's blue shirt, he thought 'I'm done for'. We ate shrimp cocktail, drank champagne and asked the recalcitrant barman to take our photograph. The problem was that these images – these beautiful, glimmering images – were not grounded in reality. The whole thing was a lie, from the taking of the photograph, which suggested an infinitely greater intimacy than did or really would ever exist between us, to the careless money we spaffed away on profligate frivolities as though, like our fellow Manhattanites, it meant nothing to us, when in fact we were each privately aware that we could not afford to live this way for long, a mutual pretence we would maintain for the whole relationship. And I, too, was a lie, or at least the version he saw and prized – the insouciant androgynous waifish woman who enchanted him – was not true, was not me. I had been extremely ill earlier in the year and lost thirty pounds and in the moment he crossed my path was still not well, but because I was thin people assumed I was healthier and happier than ever; in my temporary nimble glamour I was certainly more Manhattan than ever.

Grey was a good person. I loved him, and we shared many special things together, but I could never shake the feeling that our performance of people being enchanted by things was precisely that: an act. He had this habit of reminding me to really look at, or appreciate, something. We would be in some beautiful, romantic place, both admiring a perfect scene, and he would say, 'Isn't the sea beautiful?' and I would agree it was, and then he would say, 'No but really, look at it,' and I would think, I already am bloody looking at it. It made me feel like I was seeing the world wrong. He would gesture around us sometimes and say, in gratitude, 'Who gets to do this?' and at first I laughed in agreement, but after a while I would think, irritably, 'Who gets to do this? Not us, we don't belong here.' There was something self-conscious in our indulgence, like we didn't quite know how to pull it off without an audience. This happened one day in an overpriced spa on Governors Island where you sit outside in a jacuzzi or an infinity pool and look at the stunning unobstructed view of the city skyline. I couldn't seem to look long or well enough at it to suit Grey, and suddenly I was sick of Manhattan, of the way it demanded observation, of the way it made me feel part of a production.

Afterwards on the ferry back to Wall Street I was filled with the certainty that I had gotten something very wrong. When Grey and I broke up my perception shifted and I understood that I should never have presumed to interact with the city as anything other than a kind of tourist. Though, of course, any newly arrived transplant wants to distinguish themselves from tourists; the guilty thrill of being sincerely annoyed by dawdling tour groups blocking sidewalks for the first time. Perhaps I ought to have been treating Manhattan like a lover instead of a long-term partner.

I busied myself establishing my real life in Flatbush, next to Prospect Park, and then once I felt at home there I began to pay regular, respectful visits to the city. I was alone in Manhattan far more often than I was in company. Sometimes I would go to parties downtown or dinners uptown or to the ballet, but more often than not I would finish my work for the day in my apartment and then, in the restless mood in which I would in years past have gone on an app to find a date or somebody to sleep with, I would instead get the B train to Broadway Lafayette Street, see a movie at IFC and walk slowly and deliberately around the Village deciding where the best bowl of noodles would be. Sometimes I listened to an album by Ed Askew, an elderly musician, artist and New York legend, whose talent and niche renown had not translated to financial success, and whose medical and housing needs my friend Jay had come to assume much responsibility for, dealing with the endless bureaucracy of accessing subsidised healthcare. This filled me with admiration for Jay, who at a glance could be taken as some aesthetically pleasing downtown hipster, but in reality was a real and richly complex person, a fact I would never have known if I didn't meet him first in the spirit of gratuitous gaiety I had. Jay, Eli, John – how funny to have met them in this way so many years ago, in a way that was supposed to be all surface and no substance, and have them all still be in my life now, my great friends, people I could call if I lost everything, if I had no place to go.

I thought about that strange month in 2020 when I had met them all, when the full hedonistic possibility of the city had consumed me and I wanted everything; wanted to drink cream and spend thousands of dollars on dinner and drinks, and sleep with as many different people as I could; wanted to fuck the whole world, felt that there could be no pleasure great enough to sate me and no reason why I shouldn't get to have it all.

Now, in my solitude, it feels like the city itself is boyfriend enough, that I don't need to populate it with all the other performances. It isn't mine, it will never be a place I can claim, but it's enough, isn't it, to ride the train into Manhattan once or twice a week. Enough to have not yet lost the impulse, as the carriage trundles out of the darkness and onto the bridge and the skyline becomes visible, to stand up and peer out at it in excitement and gratitude, wanting to jostle the reading and scrolling passengers beside me and tell them: 'Look! No, really look – who gets to do this?'

About the Contributors

Anthony Anaxagorou FRSL is a British-born Cypriot poet, fiction writer, essayist and publisher. His third collection, *Heritage Aesthetics* published with Granta Poetry in 2022, won the RSL Ondaatje Prize 2023 and his second collection, *After the Formalities*, was shortlisted for the 2019 T. S. Eliot Prize. Anthony is artistic director of Out-Spoken, a monthly poetry and music night held at London's Southbank Centre, and publisher of Out-Spoken Press. He is the editor-in-chief of *Propel Magazine*, an online literary journal featuring the work of poets yet to publish a first collection and the founder and curator of *WriteBack*, a quarterly literary series held at the British Library.

ABOUT THE CONTRIBUTORS

Santanu Bhattacharya is the author of two novels, *One Small Voice* and *Deviants,* and several works of short fiction. *One Small Voice* was an Observer Best Debut Novel for 2023, and was shortlisted for the Author's Club Best First Novel Award and the Gordon Bowker Volcano Prize. Santanu is the winner of the Desmond Elliott Prize Residency, the Mo Siewcharran Prize, the Life Writing Prize, and a London Writers' Award. He grew up in India, and now lives in London.

Octavia Bright is a writer and broadcaster. She has presented programmes for BBC R4 and The World Service including *World Book Café* and *Open Book*, and hosts literary events for bookshops, publishers, and festivals, and events for The Southbank Centre. She also co-hosted *Literary Friction*, the literary podcast and NTS Radio show, for over a decade with Carrie Plitt. Her writing has been published in many places, including the *Guardian,* the *Sunday Times,* the *White Review, ELLE, Stylist, Somesuch Stories* and *Harper's Bazaar*, and in the anthology *Critical Hits*, published by Serpent's Tail. She has a PhD from UCL where she wrote about hysteria and desire in Spanish cinema. Her first book, *This Ragged Grace,* was published by Canongate in 2023.

ABOUT THE CONTRIBUTORS

Nicola Dinan grew up in Hong Kong and Kuala Lumpur and now lives in London. *Bellies*, her debut, won the Polari First Book Prize. Her second novel, *Disappoint Me*, was shortlisted for the New Adult Book Prize.

Ella Frears is a poet based in London. Her collection *Shine, Darling* (2020) was shortlisted for the T. S. Eliot Prize and the Forward Prize for Best First Collection. Her book *Goodlord* (2024), which takes the form of one long email to an estate agent, was shortlisted for The Forward Prize and a Sky Arts Award. She hosts chat and music show *Tears for Frears* on Soho Radio.

Sinéad Gleeson is the author of the essay collection Constellations and a novel, *Hagstone*. She is the editor of several anthologies, including *The Art of the Glimpse* and *This Woman's Work: Essays on Music*, co-edited with Kim Gordon. She frequently writes about art and collaborates with artists and composers.

Noreen Masud is an Associate Professor in English Literature at the University of Bristol, and an AHRC/BBC New Generation Thinker. Her academic monograph, *Stevie Smith and the Aphorism: Hard Language* (2022) won the MSA First Book Award 2023 and the University English Prize in 2024. Her memoir-travelogue, *A Flat Place* (Hamish Hamilton [Penguin]

ABOUT THE CONTRIBUTORS

and Melville House Press, 2023), was shortlisted for the Women's Prize for Non-Fiction, the Sunday Times Charlotte Aitken Trust Young Writer of the Year Award, the Jhalak Prize, the RSL Ondaatje Prize and the Books Are My Bag Readers Awards.

Orlaine McDonald is a writer of Jamaican and Irish heritage, and lives in London. Her debut novel *No Small Thing* was shortlisted for the 2024 Nero Book Awards for Debut Fiction, the 2025 RSL Ondaatje Prize and was a June Indie Book of the Month pick. She is winner of the 2025 Kate O'Brien Award.

Megan Nolan is an Irish writer based in New York. She is the author of the novels *Acts of Desperation* and *Ordinary Human Failings*, which have been translated into fourteen languages.

K Patrick is the author of *Mrs S* and *Three Births*. They live in Scotland.

Cecile Pin is a writer based in London. Her debut novel *Wandering Souls* was longlisted for the Women's Prize for Fiction, the Prix Femina étranger, and shortlisted for the Waterstones Debut Fiction Prize. She has won the Fragonard Prize for Foreign Literature, a Somerset Maugham Award, and a London Writers'

Award. In 2025, she was selected as one of Forbes' 30 Under 30 Europe. Her second novel *Celestial Lights* will be published in Spring 2026.

Alexandra Pringle was the fourth person to join Virago Press in the 1970s, eventually ending as their Editorial Director. She worked as a literary agent and at Hamish Hamilton before joining Bloomsbury Publishing, where she was editor-in-chief for over twenty years. Alexandra is a founding director of Silk Road Slippers, teaching writing masterclasses in Marrakech and London, and she's writing a memoir, *Caravan*, to be published by Canongate in the UK and Summit Books in the US in 2027.

Ralf Webb is the author of the poetry collections *Rotten Days in Late Summer* and *Highway Cottage*, as well as the non-fiction book *Strange Relations: Masculinity, Sexuality and Art in Midcentury America*, which was shortlisted for the Sunday Times Young Writer of the Year Award.

Daunt Books

Founded in 2010, Daunt Books Publishing grew out of Daunt Books, independent booksellers with shops in London and the south of England. We publish the finest writing in English and in translation, from literary fiction – novels and short stories – to narrative non-fiction, including essays and memoirs. Our modern classics list revives authors whose work has unjustly fallen out of print. In 2020 we launched Daunt Books Originals, an imprint for bold and inventive new writing.

www.dauntbookspublishing.co.uk

We ensure all our products comply with GPSR, CE marking, and other applicable EU Directives. Our EU Responsible Person for GPSR product safety compliance is EU Compliance Partner.

EU Responsible Person (EU RP):
EU Compliance Partner

Postal address: Pärnu mnt. 139b – 14, 11317 Tallinn, Estonia

Contact Email: hello@eucompliancepartner.com

Website: www.eucompliancepartner.com

Phone: +33757690241